Praise for *Bits, Bytes, and Barrels*

"Peak oil, scarcity, and rising prices? No, the energy industry is in a new era of abundance and price pressure. In *Bits, Bytes, and Barrels,* the authors give valuable context and advice on how digital technologies are mandatory tools for progressive energy leaders to turn disruptive challenges into a generational opportunity."

PETER TERTZAKIAN, executive director, ARC Energy Research Institute

"*Bits, Bytes, and Barrels* comprehensively lays out the issues and opportunities for digital technologies in the energy business. This book is full of examples where digital is making a bottom line impact for companies and it provides pathways for how companies might proceed in their own digital strategies. The authors have taken a very technical and complex subject and made it clear and understandable—a great handbook for executives and IT professionals alike!"

BRIAN TRUELOVE, former SVP and CIO, Hess Corporation

"Technology, innovation, and business model changes are impacting industry at a rapid pace—yet in some cases, not fast enough. *Bits, Bytes, and Barrels* is an excellent primer on existing and future innovation levers that the energy sector can incorporate to improve adaptability and competitiveness. Important reading for Boards, the C-suite, and those looking to understand how technology can move the needle or stand-up effective internal innovation. I highly recommend this book."

SAMANTHA STUART, vice president, strategy and corporate development, TransCanada

"*Bits, Bytes, and Barrels* offers a candid discussion of not only the potential uses for digital, but also the possible pitfalls of poorly implemented projects. The numerous examples explored are enlightening and offer proof points of why companies should act. It is a welcomed perspective that should be helpful to those who are considering greater deployment of digital technology in their organizations."

TOM MUECKE, retired after forty years in upstream research and operations at a US supermajor

"This is a timely, well-researched, practical, and insightful book that plugs a much-needed gap in the market. If you are looking for a one-stop shop that brings together key digital themes and ideas in oil and gas in one place, then this is it—highly recommended."

DR. JOHN PILLAY, global digital transformation director, WorleyParsons

"This practical, easy-to-read book provides a guide to the current state of digital innovation in the oil and gas industry and where to expect the leaders to take it next. A must-read to ensure your oil and gas firm stays competitive."

JUDY FAIRBURN, Board director, former EVP and chief digital officer for a large energy firm

"It's interesting that oil and gas come from dinosaurs, because 'dinosaurs' describe our industry when it comes to the adoption and implementation of technology. Cann and Goydan serve a satisfying menu of bite-size topics in their book, *Bits, Bytes, and Barrels: The Digital Transformation of Oil and Gas.* As a downstream company in the (dinosaur) industry that is slowly implementing a digital transformation within our culture and plant, I agree with the authors that taking the first step is the hardest thing to do—but once that first step is taken, the next ones are easier. In this book, they make it clear that as an industry, we either embrace digital change, or get crushed by it. Anyone in our industry who is interested in not becoming extinct should read this."

DOUGLAS H. SMITH, CEO, Texmark

"This new book is an essential road map for oil and gas executives and Boards to navigate the new world of digital enhancement of their business, avoid false starts and missteps, and unleash new sources of value. This book brings much-needed clarity to a topic which can be confusing; and makes it relevant to the real world of oil and gas at every step."

ANDREW SLAUGHTER, executive director, Deloitte Center for Energy Solutions

"Geoffrey Cann and Rachael Goydan provide a highly accessible snapshot of the complex and fast-moving digital world as it applies to oil and gas. Many oil and gas workers feel this is an area in which it is all too easy to let new ideas and technologies slip by. This book provides with reassuring confidence a contemporary, forward-looking state of the digital world that will allow oil and gas executives, Board members, engineers, business developers, researchers, and strategists to assess areas to develop and focus, and also reveal blind spots."

ANDY DOYLE-LINDEN, consultant and corporate strategy lecturer, EMBA, Haskayne School of Business

"For many of us in the oil and gas industry, the path from the old way of doing business into the new frontier of the digital world can be daunting and confusing. It's difficult to find practical and actionable advice and insights on how to best make this transition. Rachael and Geoffrey have broken this journey down into understandable steps and provide a clear and thoughtful way for us to make this important change."

MUSHAHID KHAN, CEO and Board member, APS Plastics and Manufacturing

"The perspectives in *Bits, Bytes, and Barrels* define the white spaces that people are considering when they think of the digital oil field. It validates many key points on the topic of what digital transformation is, and provides true insight for those seeking some reasonable understanding of how hydrocarbon development needs to be digitized as the demand for oil continues to increase."

DAVID SMETHURST, chief technology officer, energy sector, Hitachi Vantara

"Digital solutions will inevitably transform the energy sector. *Bits, Bytes, and Barrels* provides an excellent road map for energy executives to get started on the digital path. Geoffrey and Rachael bring a global perspective, drawing on successful models from around the world and from other sectors, along with pragmatic advice on how to engage with emerging technologies. I highly recommend this book for anyone involved in the energy sector."

MARTY REED, CEO, Evok Innovations

"As a petrochemical manufacturer and an early adopter of IIoT solutions, I wish this book were published before Texmark and our partners began our digital journey two years ago. It was encouraging to read stories about solutions from around the upstream world and inspiring to read about what's possible. Digital solutions and innovative mindsets have completely changed the way our lean, diverse team of operators, millwrights, engineers, and technical staff approach new business, existing process optimization, safety, troubleshooting, training, and daily operations. I recommend this book to leaders looking for the courage to take a baby step toward employing digital solutions... and then I recommend they give a copy of it to everyone in their organizations."

LINDA SALINAS, vice president of operations, Texmark

"Geoffrey Cann and Rachael Goydan's *Bits, Bytes, and Barrels: The Digital Transformation of Oil and Gas* is a handbook of how to develop digital strategy and culture and get IT and operations engaged. This book is both interesting for those familiar with oil and gas, and easy to understand for those who are not. A great resource."

SOHEIL ASGARPOUR, CEO, Petroleum Technology Alliance of Canada

"*Bits, Bytes, and Barrels* is a balanced critique of the industrial digital technologies which can feel like a bewildering maze for organizations facing the digital transformation challenge. A must-read for energy and mining executives and Board directors seeking to set their organizations' strategic path towards a connected future."

AZAD HESSAMODINI, president, strategy and development, Wood

Bits, Bytes, and Barrels

GEOFFREY CANN AND
RACHAEL GOYDAN

BITS, BYTES,

AND BARRELS

The Digital Transformation of Oil and Gas

ISBN 978-1-9995149-0-7 (paperback)
ISBN 978-1-9995149-1-4 (ebook)

Produced by Page Two
www.pagetwostrategies.com

Cover and interior design by Taysia Louie

geoffreycann.com

We dedicate this book to:

Marjorie, for her unwavering faith and support throughout this journey, and for listening to Geoffrey read every single word aloud

Sampat, for his integrity, dedication, friendship, sage advice, and steadfast sponsorship throughout Rachael's career

CONTENTS

INTRODUCTION

THE OIL AND gas industry is well known for its boom-and-bust cycle—as the wells peter out, demand continues to rise, ensuring the next boom. In 2008, prices collapsed from well over $100 per barrel (bbl.) to as low as $30, only to resume the climb to $100/bbl. again by 2014, as demand grew to 95 million barrels per day.

But now the global oil and gas industry is truly at a crossroads. For the first time in many decades, the confluence of a number of forces is flattening out the growth in petroleum, particularly for transportation. Rapidly advancing alternatives in drivetrain technology, including hybrid and electric motors, dramatically better renewable energy sources and battery technologies; shifting consumer preferences towards shared vehicles; and a global movement to decarbonize look poised to exert unrelenting downward pressure.

These changes vary immensely from country to country, with some smaller European nations aiming to fully eliminate fossil fuels from their power mix. China has explicitly set out in its latest five-year plan its intent to be a global leader in renewable energy, battery technology, and new transportation technologies, while still presenting a strong growth prospect for petroleum consumption. Other large nations, such as India, also provide ample opportunity for growth.

For many oil producers, the standard playbook for managing the down-cycle of the commodity market has run its course. Procurement teams have mercilessly squeezed the supply chain and extracted painful price concessions. Management teams have high-graded capital budgets and now only chase the best opportunities. Several hundred thousand jobs have been eliminated from the industry, mostly from the ranks of the service companies and engineering firms. Bankruptcies have peaked, and boards have replaced their CEOs. Cost-reduction efforts have now brought the industry to a level of profitability that exceeds the level when oil prices peaked in 2014.

ROYAL DUTCH SHELL'S reported 2018 first quarter results show the company making as much profit at $60 per barrel as it did when oil was $100 per barrel. Goldman Sachs also reported in early 2017 that European oil companies generated the same amount of cash at $52 per barrel as they did at $109.

Meanwhile, actual and potential supplies of hydrocarbons are rising at a remarkable rate. The industry has successfully adopted technological innovations that have caused a permanent swing from peak oil to super abundance. In response, many global producers in 2015 and 2016 set out to maximize their production to maintain market share. Prices have dropped from historic highs of more than $100/bbl. to record lows of $28, only to slowly climb back up. As I write this passage in mid-2018, the price of oil is still just $67/bbl. Indeed, some industry watchers have concluded that we have discovered far too many fossil fuel resources to be consumed while remaining within the 2° threshold agreed upon in the Paris Agreement on climate change.

Some cities and nations are acting now to address climate change and fossil fuels' environmental impact. Britain, France, Germany, the Netherlands, Norway, and several others have announced bans on gasoline and diesel vehicles to take effect in the years to come, some as soon as 2025. France and New Zealand plan to end oil and gas production altogether. Norway's sovereign wealth fund, the world's largest which was built almost completely on its oil and gas exports, intends to divest from holdings in fossil fuel businesses.

Digital innovation is playing a decisive role in the fortunes of oil and gas as a driver of three big changes:

1. it is expanding the supply of hydrocarbons,
2. it is lowering cost and increasing productivity, and
3. it is helping to erode demand for fossil fuels.

Artificial intelligence and machine learning expand supply by unlocking vast new troves of low-cost, high-quality petroleum, particularly in the new shale resources. Costs fall as drones and submersibles carry out the more dangerous and costly work, which would otherwise involve humans, in the most remote places where oil and gas are found and produced. Productivity improves as analytics can help optimize production assets and flag emission occurrences. Business efficiency, through blockchain solutions, reduces the cost of counting and transacting the movement of hydrocarbons throughout the value chain. Connected, shared, and autonomous vehicles reduce fuel demand, and the adoption of all of these same technologies by many other industries in their supply chains also lowers fuel consumption.

In addition to potentially increasing supply, improving cost, and eroding demand, digital innovation can also enable new business models with the potential to disrupt the oil and gas industry more profoundly. Gasoline customers can order fuel to be delivered to their vehicles via apps, reducing customer visits to fuel stations and potentially stranding an entire asset class. Cloud computing enables industry participants to pool their resource information and access best available interpretation services in ways that have been only available to the largest players. The separation of the data about resources from the resources themselves potentially creates new asset-light market entrants.

The oil and gas industry has always been an aggressive user of computer technology, particularly in exploration, where it operates some of the world's fastest supercomputers. But companies have only tentatively embraced the wave of digital change that has revolutionized other sectors such as retail, entertainment, and financial services. A go-slow approach to digital has disrupted established business models in many industries already (i.e., media, music, retail, and consumer goods) while creating enormous wealth for innovators elsewhere.

ACCORDING TO STATISTA'S global ranking reports, for 2017, six of the ten largest companies in the world by market capitalization are all digital—Amazon, Apple, Facebook, Alphabet, Microsoft, and Tencent. *Forbes* reports that in 2017, four of the top ten largest companies in the world by revenue were oil and gas companies—China National Petroleum, Sinopec, Royal Dutch Shell, and ExxonMobil. Only ExxonMobil and Apple are on both top ten lists.

Oil and gas is generally cautious when it comes to embracing change, and for good reasons—safety being a primary one. However, in this case, the industry must shed its risk aversion and embrace change, or risk being lumped in with other pariah industries, such as asbestos and tobacco, with dire consequences for its ability to attract and retain the next generation of knowledge workers. Indeed, many industry executives are already asking how they can transform their business at the speed of digital change that they see in so many other sectors.

This book explores a number of questions related to the digitization of oil and gas, including:

- What is digital?
- What are the most important digital technologies to consider?
- What is the business case for digitization in oil and gas?
- Why is it essential for the industry to embrace digital now?
- How will the profile of talent in the industry change?
- What are the oil and gas industry's specific risks and barriers to adoption?
- What approaches to adopting digital change would be most beneficial?
- How can a company get started with digitization?

Some will think it a Herculean task to marry up the largest industry by market capitalization with one of the largest industries by revenue. Both industries are enormous in scale and scope. Others will consider it more of a Sisyphean task, especially since the digital industry is changing at an exponential rate. In any event, I anticipate a need to refresh the book's findings in a few years' time.

While aiming primarily at the oil and gas industry, I set out to write this book with several goals in mind, namely to:

1. **Inform the conversation.** Industry leaders are being bombarded with media stories, conference opportunities, and pitches from technology companies about digital. Technological innovation has moved from the domain of a small number of industry insiders and incumbents to a huge range of startups.

2. **Embrace the whole industry.** The oil and gas industry is broad—onshore, offshore, upstream, midstream, refining, distribution, and retail. Digital innovation impacts the breadth of the industry and is not confined to consumer-facing retail.

3. **Remove bias.** Most publications on the topic of digital innovation in oil and gas are created by technology companies or business consultancies. These publications lack independence, which in part impedes the industry from acting.

4. **Speak plainly.** The digital industry and the oil and gas industry have their respective terminology, abbreviations, and insider language. To be effective, the conversation needs to cut through the obscurity and provide meaningful insight. That said, the use of some industry terminology is necessary, and to aid understanding, there is a glossary provided at the back of this book.

With this in mind, the book will appeal to many different audiences:

- Boards in oil and gas, who may be struggling to understand what digital is, its threats, and its opportunities.
- Management teams in oil and gas, who are tasked with identifying digital opportunities and implementing change in their organizations.

- Technology companies, who have clever solutions that could be valuable to the industry, but may not understand the sector and its nuances.
- Regulators and policy makers, who need to stay abreast of how technologies can make positive impacts on society, the economy, and the environment.

With such a complex agenda, how should you approach this book? I suggest all readers begin with Chapter 1, which provides definitions for digital that underpin the rest of the book. Chapter 2 explores the handful of key digital technologies that are likely to have the greatest impact on the bulk of the industry, and it should appeal to everyone, including technology companies. Chapter 3 looks at digital changes that are specific to different segments of the industry; I would be selective and read only those sections that match your business. In Chapter 4, I review the key challenges to overcome for digital benefits to be captured, including talent and skill shortages, cyber risks, the cautious business culture of oil and gas, and institutional impediments. Finally, Chapter 5 presents a road map for getting started and accelerating digital adoption in oil and gas. Each chapter ends with a few key messages to take away and consider in your specific context.

And when you are done, regardless of your role in business or government, you'll have a better understanding of how digital technologies have the potential to fundamentally transform oil and gas for the better, and how to successfully deploy digital solutions in practice.

You may have noted that this book has two authors, but for simplicity's sake we have written it from a first-person perspective.

Geoffrey Cann

I am a business advisor to the oil and gas industry, a role I have fulfilled for more than thirty years. After graduating from McGill University with an undergraduate business degree in computer programming, I landed at Imperial Oil in Toronto, where I began to appreciate the scale, scope, and impact of the oil and gas industry on society, and the role of technology in the industry. Eventually, I returned to the Ivey Business School for an MBA and then joined Deloitte, the world's largest professional services company. I have since worked up and down the oil and gas sector on multiple continents and settings, advising such organizations as Suncor, Husky Energy, Canadian Natural Resources (CNRL), Enbridge, Irving Oil, Origin Energy, Precision Drilling, Calfrac Well Services, CE Franklin (now called DistributionNOW), Korea National Oil Corporation (KNOC), and many others. Since retiring from Deloitte, I continue to advise Boards and management teams, in both the oil and gas industry and in the technology industry, on the impacts of digital innovation on the industry.

Rachael Goydan

I value lifelong learning and innovation and always aim to integrate these concepts into the work I do. Since joining Deloitte in 2001, I have worked as a consultant with clients across multiple industries. My journey has taken me to work directly, or as Board volunteer, in education, banking, telecommunications—and for the last twelve years—energy. I eventually met Geoffrey when we began collaborating to bringing digital innovation to clients across the energy value chain. I earned my undergraduate degree in psychology from the University of Pennsylvania and my MBA (with a focus on information technology and business transformation) from the Sloan School of Management at MIT.

1

WHAT IS "DIGITAL"?

Data, Analytics, and Connectivity

DESPITE THE NEAR-UNIVERSAL use of the term "digital" to represent features of our modern economy, there is no shared definition of what digital actually is. To an engineer, digital could be the opposite of analog. To a millennial, digital is an integral part of everyday life. To a lawyer, digital could be a new kind of asset that has different intellectual property rights. To a banker, digital is the way of business in the future.

I am constantly asked to define digital, and frankly, it's devilishly hard. But my best examples include three key building blocks that together create something that most people generally concede is a digital device, solution, or service.

- Data—data is the lifeblood of digital. A digital device, solution, or service produces and uses data.
- Analytics—a digital device, solution, or service has the ability to carry out calculations and computations on the data.
- Connectivity—a digital device, solution, or service uses a telecommunications network that allow digital devices to connect with one another to exchange or share data, or computations.

Something that is digital (and that something could be a physical thing, a process, or a business model) has these three basic elements operating together in some configuration.

A smartphone is an excellent example of a digital thing: it has data, such as address books, music files, and maps; analytics, which are apps that carry out calculations, such as the distance between two points; and connectivity, since a phone, by definition, can use the cell phone network and likely has multiple network connection technologies embedded within, including Wi-Fi and Bluetooth.

Other examples of digital things in the oil and gas world include:

- Tank gauges—Fuel tank gauges have shrunk in size, cost, and power demand while expanding in capability. Australia's unmanned outback airports have gauges on their tanks that give fuel providers real-time visibility to tank contents, so that the tanks can be replenished when needed.
- Cars—Next-generation vehicles are packed with digital smarts to allow them to communicate with each other and with smart transportation environments. Porsche is embedding blockchain technology in its sports cars.
- Valves—With sensors and actuators falling in cost, even traditionally dumb devices like valves can be brought online, generate their own data feed, and tie into supervisory systems. The same for drill bits, flow-measurement devices, motors, and filters (the basic building blocks of process manufacturing).

JUST ABOUT ANYTHING can become digital and participate in the digital world. Here's a perhaps silly example, but one that's instructive: a fish tank, albeit a high-tech one with lots of sensors to monitor temperature, salinity, lighting, and pH balance, included a network connection that sent the data the sensors collected to a remote monitoring system. A digital fish feeder was recently exploited by criminals to steal log-in IDs and passwords from a casino.

These three building blocks (data, analytics, and connectivity) share a common foundational technology, which is the lowly computer chip. As chip technology advances, the cost of earlier generations of chips falls to zero (manufacturers basically give them away) and it becomes economically feasible to incorporate chips into almost everything. They become smaller, thinner, lighter, and, most critically, they need very little power to operate. It is the chips that store the data, provide the computations, and enable the telecommunications computers that drive the network connections.

We've all read about the exponential growth curve of transistors and now computer chips and about this growth's impact on the electronics industry (see sidebar). It turns out that the three components that make something digital (data, analytics, and communications) are each experiencing massive expansion because they are themselves based on computer chips.

GORDON MOORE, THE co-founder of leading computer-chip maker Intel, noted with curiosity in 1965 that the density of transistors on a computer chip appeared to double every year. This rate, a consistent doubling per unit time, is called Moore's law. Things that demonstrate this kind of exponential growth rate early in their life cycle bear watching as they appear to suddenly become very large without notice. Many things in the digital world share this phenomenon.

Occurring in parallel with the exponential growth rate in capability is an exponential fall in cost. At some point, the capability of a digital thing is effectively infinite in terms of data storage and processing—and practically free as the cost approaches zero. Where the oil and gas industry has always worked in a world of constraint, digital is creating a world of abundance.

I first began to notice the impact of digital on my own life when I was traveling on business. Fifteen years ago, the sole electronics that I might have packed for travel included a laptop and portable phone. Today I travel with a jumble of electronics, including a smartphone, a smart watch, digital pens, a tablet, the company-issued work laptop, mouse, wireless headphones, wireless earbuds, spare rechargeable batteries, plus all the cabling and outlet jacks. Rarely do hotels offer anything close to the number of outlets I need to recharge everything. Soon, articles of clothing and luggage will be electronic and the hunt for hotel room electrical outlets will become even more pressing.

THE DATA TORRENT

The volume of data our societies create, represented as zeros and ones in a storage device, have been growing by truly staggering amounts. According to the International Energy Agency (IEA), IBM estimates that between 2015 and 2016, the world generated almost as much data (90 percent) as already existed in all of the world's storage systems. Our devices and our increasingly digital lives contribute to this growth rate in many ways.

- A high-quality photo comprises eight to ten megabytes of data. Most of us do not even discard photos anymore. They simply pile up on our phones and home computers.
- A high-quality ten-minute video taken on our smartphone takes up 1.5 gigabytes. Four hundred hours of video are uploaded every minute to YouTube.
- A typical flight produces a terabyte of data, and autonomous vehicles and trucks will generate similar data volumes.

Industrial data volumes are not growing at quite the same pace as consumer data, but that's because industry has not yet equipped all its various assets, tools, and people with sensors. But as industrial assets become data generators, they will match the prolific data generation of humans. An Airbus or Boeing aircraft is festooned with hundreds of sensors that produce discrete data measurements every tenth, hundredth, or thousandth of a second throughout a flight.

A NEW REFINERY, Canada's first in a generation, rises from the industrial land northeast of Edmonton and will incorporate 25,000 sensors, an order of magnitude more than its predecessors. Each sensor will generate a steady flow of measurement data every second. The future of digital plants has arrived in oil and gas.

Beyond the growth in volume, data is also changing shape. Early generations of computer systems could only process highly structured data, such as rows and columns on a spreadsheet, in the form of numbers and letters. But modern data can take almost any shape, including unstructured data like photographs, waves, sounds, video, and sensations like vibrations and smells.

The DNA of living things can be represented as data, a very important development. Industry will soon be able to tell the provenance of a thing or substance by investigating its DNA and comparing it to a registry of DNA samples. The oil and gas industry is not far from a future where it will be able to take a sample of a barrel of oil and identify which oil basin it came from by testing microbes found in oil samples (although perhaps not after crude oils are blended in pipelines and tanks). Ethical producers of oil, those nations that invest more to protect the environment, will be able to command a premium for their product.

An emerging challenge is how to analyze this flood of data. Tools and techniques for data analysis, interpretation, visualization, and monitoring need to keep pace with the growth in data volumes, the flow of that data from one location to another, and the analytic demands of data users. The spreadsheets of the past are simply not up to the task.

CHEAP MATH

Analytics, a term for computer horsepower, is also demonstrating the same rapid development and growth as data, which isn't surprising because analytics are carried out on computer chips, too (though not the same chips that store data). Think of "analytics" as consisting of these math computer chips mounted on circuit boards and the software that carries out the math. Both data and math chips have been advancing with vigor, and along important lines:

- Chips can be both highly specialized (for such intense tasks as video gaming or robot controlling) and generalized (for laptop and desktop computers). Car makers now use video-gaming chips' ability to process visual data as the basis for autonomous car navigation systems.
- Sensors are basically chips. Inexpensive sensors for GPS, orientation, cameras, light, sound, and speech-processing are all just math chips with embedded software.
- Chips need power to operate—to keep them running longer and more continuously. To prevent them from overheating and depleting their batteries, engineers have been designing them to be power-stingy, and power consumption has been falling with each generation of chip.
- Along with their power profile, physical size, and capabilities, their costs have fallen dramatically, allowing chips to be installed in many unexpected devices including pumps, valves, and gauges.

We can see the impacts of these developments on a personal level. My Apple Watch, which retails for a few hundred dollars, has the equivalent analytic capability to a multi-million-dollar Cray supercomputer from the 1980s. A smartphone has much the same capability of the mainframe systems that enabled NASA's moon-shot programs in the 1960s, and, in just ten years, smartphones advanced from a novelty to a must-have for modern life.

Industries have varied in their drive to put sensors onto things, but the pace of adoption is accelerating. Roland Berger estimates that 15 billion smart sensors were installed globally in 2015, but by 2020, the number of sensors is expected to be greater than 30 billion. This phenomenon is giving rise to the Industrial Internet of Things (IIoT)—pumps and motors, valves and gauges, pipes and tanks, wires and switches are being fitted and retrofitted with sets of sensors. Once in place and live, the sensor adds to the flood of data, draws on the power grid, and creates demand for connectivity to send that data elsewhere.

The prize for industry is enormous. Greater visibility through sensors means that health and safety issues, like leaks and spills, can be detected faster, even before they occur. Better understanding of equipment performance means equipment can run longer and at higher rates without failure. Faster collection and processing of more accurate field data means faster payment by oil companies for the services they purchase.

Chips that can perform calculations need software code to execute the math. Similar to the chips, software development shows the same exponential growth characteristics.

- Programming languages are becoming easier to learn. They feature drag-and-drop interfaces, high reusability, and standardization so that coders are more productive. Compute needs like sorting, analysis, and mapping are simple and built-in.
- Programming languages are becoming more ubiquitous and common to multiple uses, rather than specialized by asset or application. Recruiting coders is much easier.
- The use of common chip sets means the development of industrial software for a piece of equipment is not dramatically different from coding for a commercial system or a web application.
- It is now acceptable for companies to rely on open-source code—software that is developed freely and shared openly—which both accelerates the delivery of new solutions to business and lowers the cost.

BIG PIPES

The final building block in a digital world is connectivity. Without connectivity, a device that has data and analytics is just a calculator and, these days, not a very useful calculator.

Low-cost chips, analytics, and software development have helped transform the telecommunications sector in just one human lifetime to enable extraordinary connectivity. From 1991, with the launch of 2G, the telecom industry has progressed to 3G, 4G, LTE, NFC, Bluetooth, and now 5G.

As a result, the world is becoming highly interconnected. According to the IEA, at the dawn of 2016:

- the number of households globally with Internet service was around 54 percent, compared to 80 percent that had access to electricity, concentrated among developed nations, but accelerating in the developing world;
- the number of individual Internet users was about 3.5 billion;
- the number of mobile phone subscriptions, a measure of the number of users able to tap into digital services, reached 7.7 billion; and
- the number of inanimate objects or things that are connected to the Internet is estimated to be about 8 billion and will grow to 30 billion by 2030.

The volume of data that networks move provides a good indication of the penetration of and demand for connectivity. In 1974, the total amount of data that was transmitted on worldwide networks in a month was one terabyte. By 2016, worldwide networks move one terabyte every second, an increase of 2.5 million times.

There are no signs that this volume of data is flattening out. If anything, the growth in the number of sensors and the greater capability of analytics suggests that data volumes are likely to continue to grow. Indeed, certain technical innovations in our digital world are still very early in their own adoption life cycle (such as autonomous vehicles), suggesting that the demand for data connectivity has serious propulsion.

CANADIAN NATURAL RESOURCES, a leading oil producer, had entered into an auction for telecommunications spectrum in competition with the nation's wireless providers. Effective telecommunications have become a priority for large industrial concerns who need better networking capabilities for the future.

One clear issue for the oil and gas industry, even in very developed nations like Canada, and still very pressing in developing nations, is the lack of robust and ubiquitous telecommunications infrastructure. Oil and gas is often found far from concentrated areas of civilization, and telecom companies have been slow to roll out network connectivity. This continues to plague both Australia and Canada, as well as the offshore industry in general. Telecom technology still suffers from intermittency during weather events, a serious challenge for dangerous infrastructure that needs constant supervision.

LESSONS FROM THE FRONT

While the digital wave of change has had only a modest effect on oil and gas to date, early adopters in other industries have been much more profoundly impacted. Several lessons about digital warrant consideration for later adopters, such as oil and gas, in order to avoid the same pitfalls and to take full advantage of digital's might.

New Business Models

The most important insight is that digital solutions unlock new and different business models, where value is exchanged in exciting new ways. For example, PlantMiner aggregates spare capacity of earth-moving equipment and makes that capacity available to a global market; it's one of several similar services. Cars2Go can turn virtually any parking spot into a rental car kiosk, eliminating the tediousness of rental-car contracting. Whim, a startup in Finland, is pioneering the idea of mobility as a service where its customers have unlimited access to rental cars, shared cars, taxis, buses, ferries, Uber, and rental bikes for a low monthly subscription fee. GasNinjas is an app that allows car owners to purchase fuel and have it delivered to their vehicle; without owning fuel stations, GasNinjas is in the fuel retail business and can even outsource the fuel delivery truck.

What's fascinating about many of these business models is how they tend to be asset-light and information-intense, placing more value on data than on infrastructure. Many in the world of commodities and heavy industry still believe pretty profoundly that asset ownership and ownership of all asset-related data is the key to economic success. At some stage, the risk is that a new entrant will figure out how to run a heavy industry company without owning the assets.

AN EMERGING SERVICE in the oil and gas industry is called Digital Oil Recovery: an oil company provides some of its underperforming assets to a more digitally savvy operator who then applies the latest in digital thinking to the assets. Both win by sharing the improved production, but the digital service company gains more from the experience. It is not a stretch to see how such a digitally advanced company may become the Uber for the production industry—all the software, but none of the assets.

The Importance of Ecosystems

With the digital world evolving in so many dimensions at the same time and at a pace vastly faster than most organizations can absorb, players quickly conclude that they cannot do it all, and certainly not at a level of excellence. Digital entrants tend to specialize in some niche area to achieve early cash flow and some market position and dominance. Ecosystems of connected organizations tend to emerge, with a few large anchor companies and a halo of smaller specialist outfits.

Ecosystems are everywhere in the digital world. For example, much of this book was written using an Apple iPad Pro, and at one time I thought almost the entirety of the iPad's capabilities were the product of the talented team in Cupertino. But a quick look at the legal notices on the iPad reveals dozens and dozens of companies whose intellectual properties have contributed to Apple products. Apple clearly recognizes that it does not have a monopoly on good ideas and innovation, and so nurtures an ecosystem of companies whose technologies can play a role in the Apple product line.

Successful companies that pursue digital seem to do so in the context of a successful ecosystem, either as large anchor players, like Apple, or as technology specialists. Companies that choose to go it alone are at a disadvantage.

The Network Effect

The telephone introduced us to the network effect, a phenomenon whereby the bigger the network, in terms of nodes or nodal points, the greater the value of being on a network. Having a phone is of little value if you have no one to call, but once everyone has a phone, being on the network becomes indispensable. Further, the cost to add one more node becomes practically zero.

Digital innovators design solutions in anticipation of the network effect, to achieve maximum participation as quickly as possible by all those connected phones, users, and devices. Designers use existing cloud services to avoid creating their own compute infrastructure. Software gets distributed using app stores, not in-house CD burners. The user interface is so simple to use and intuitive that training is not offered and not required. Software is often open source so that it is freely available to anyone to incorporate into their own digital innovation and it replicates quickly across solutions.

The network effect has other benefits. Large networks deliver scale, reduce costs, and improve the productivity of its users. Innovative enhancements can be quickly and cheaply delivered. Value-added services (yesterday, the Yellow Pages; today, search, trending, and analytics) become revenue generators. The bigger the network, the greater the value of these attributes.

Speed to Market

Digital solutions demonstrate an entirely different pace of adoption than their non-digital counterparts. At the time of Edison, the rollout of the phone system took years. Few companies had the scale of operations or the balance sheets to build out national or global systems quickly. Even hiring enough workers to build a physical network system required an enormous effort.

These constraints have been largely removed for digital innovations that leverage ubiquitous networks, billions of smartphones, and cloud computing. Now, almost anyone with a good idea can build a global solution at breathtaking speed. For example, recent new rideshare services launched in eighty-four countries within twenty-four months. Tesla, reacting to a complaint from a customer that cars were using plug-in stations for all-day parking, coded a feature that limited time on a plug to just the duration it took to recharge the battery. This feature was implemented in a week to all Tesla charge points.

Today, a specific goal for digital innovation and innovators is to anticipate the benefits and impact of the network effect, as well as to design their solutions to encourage hyper growth and rapid take-up.

Cyber Worries

Not a week goes by without a media story about the latest cybercrime. From ransomware to denial of service to data theft, it's clear that these digital technologies both create a greater attack surface for bad actors to exploit and are being deployed themselves as bots to disrupt industry. Companies that sell industrial equipment may be eager to embellish their valves, pumps, and motors with digital smarts, but their equipment must allow for patches to repel the latest virus. They need to allow for remote control so that they can be switched off at the first sign of attack. They need to withstand being highjacked for Bitcoin mining or other schemes for theft of energy and processing cycles. Take a lesson from the early digital adopters who have underestimated how the nefarious will promulgate fake news, disrupt elections, and create needless social discord and dissent. Cyber disruption is such a critical issue for the industry that I will devote a portion of Chapter 4 just to this topic.

Follow the Money

Digital industry leaders excel at piecing together its components (data, analytics, and connectivity), exploiting the lessons from early adopters (the network effect, speed to market, falling costs, and new business models), and using these tools to cut across business silos and achieve outsized growth rates. At the time of writing, the largest companies in the world, as measured by market capitalization, are the digital brands that have grown over the past fifteen years into colossal global companies. They have handily displaced the largest global companies of the past, many of which were oil and gas companies, in market value.

Valuing Asset-Heavy vs. Digital Companies

MARKET CAPITALIZATION

2017	Apple	Alphabet	Microsoft	Amazon	Facebook
2012	Apple	ExxonMobil	CNPC	Microsoft	Walmart
2007	ExxonMobil	GE	Microsoft	Royal Dutch Shell	AT&T
2002	Microsoft	ExxonMobil	GE	Walmart	Pfizer

REVENUE

2017	Walmart	State Grid	Sinopec	CNPC	Toyota
2012	Royal Dutch Shell	ExxonMobil	Walmart	BP	Sinopec
2007	Walmart	ExxonMobil	Royal Dutch Shell	BP	Toyota
2002	Walmart	ExxonMobil	GM	BP	Ford

DIGITAL COMPANY ENERGY COMPANY OTHER COMPANY

- Amazon, a pioneer in developing online shopping, is now a leading player in e-books, video and music services, reseller and logistics, advertising, cloud computing, and entertainment, and has big investments in artificial intelligence and machine learning.

- Apple, the famous designer, manufacturer, and marketer of mobile phones, computers, and music players, also offers digital services including iTunes, the App Store, Apple Pay, and Apple TV.

- Alphabet, the market leader in online search and advertising, offers services in fast fiber, the Internet of Things, and medical devices, as well as possesses a huge range of online assets like YouTube, Maps, and Gmail. Google is fast becoming a leader in cloud computing and artificial intelligence.

- Facebook began as the world's largest social media site and is now a major news, advertising, and shopping site, with additional services in mobile messaging (replacing telephone calls) and photo-sharing services.

- Microsoft, developer of the original personal computer software and operating system, is today a leader in cloud computing, database, enterprise resource planning (ERP), collaboration and productivity tools, gaming platforms, search engines, and social media.

- Tencent is the market leader in Chinese social networks and gaming, online music distribution, videos, and entertainment. Its WeChat app has almost a billion users.

Any industry targeted by these leaders, either directly (like retail through online shopping) or indirectly (like phone services through chat), can find itself under siege. The business press is full of stories of industries, suppliers, and corporate incumbents that did not grasp the significance of these and other digital innovators and struggled to adapt fast enough. Sectors particularly hard hit include retail, from online shopping; real estate for shopping malls, which house the retailers; the music and TV industries, from streaming services; publishing, where books, magazines, and newspapers have struggled; photography, particularly the film industry; the game industry, where video games double the sales of board games; and the taxi industry, from rideshare platforms like Uber.

Entrepreneurs now seek out investment opportunities where digital creates the potential for exponential growth rates, rapidly declining costs, and network effects: once something starts growing that quickly, it becomes very difficult to displace.

A quick scan of published reports hints at where large digital companies are investing for the future:

- autonomous cars and transportation services;
- artificial intelligence, machine learning, and robots;
- voice, interpretation, and translation;
- electrical power generation;
- residential automation;
- wearable technology, such as Google Glass or smart watches;
- business-to-business sales;
- medical information;
- financial services and banking;
- healthcare technology; and
- games, entertainment, and experiences.

My survey suggests that digital companies are not targeting the oil and gas industry directly but focusing on key sectors vitally important to the industry—notably transportation. Connected, shared, and autonomous cars could alter the demand for gasoline and diesel permanently, but more important, they create new disruptive business models. Many of these same innovations (artificial intelligence, wearables, financial services) could be put to work in oil and gas to lower costs, raise productivity, and enhance sustainability, possibly unlocking disruptive value. In fact, many companies are already experimenting with digital to do exactly that, which is the topic I will turn to next.

KEY MESSAGES

Here are a few key takeaways from this survey of the rapidly developing world of digital innovation:

1. Digital is made up of data, analytics, and connectivity.
2. Virtually anything (services, products, and assets) can and will be digitally enabled because of the falling cost of chip technology.
3. The building blocks of digital are also in hyper-growth mode.
4. Early successes in industries that adopted digital focused on building ecosystems, concentrating on speed to market, and enabling new disruptive business models.
5. Capital markets value digital companies with more optimism than asset-heavy and revenue-rich companies.
6. The market leaders have perfected their skills at bringing digital innovation to industry sectors and are investing to bring their know-how to other industries, including energy and transportation.

ESSENTIAL BUILDING BLOCKS: The Digital Tools for Oil and Gas

“Digital is just a campaign by technology marketing to embarrass oil and gas people into spending money. I’m not going to fall for it.”

A CIO OF A LARGE MULTINATIONAL UPSTREAM OIL AND GAS PRODUCER

NOT ALL DIGITAL technologies are equal: they are at different stages of maturity and have different kinds of impacts. They are not uniformly valuable. For example, ERP systems, such as SAP and Oracle, are essential to the management models of most large oil and gas companies and have been widely adopted since the turn of the century. These systems are reacting to and adopting the same underlying digital innovations and are highly mature in their fit with oil and gas. Digital twin modeling is very useful for complex networked assets but much less so for individual retail stations. Other technologies, like blockchain, are much less mature but will also have a profound impact on the industry.

The oil and gas industry has its share of skeptics, and deservedly so. Many innovations brought to the industry fail to deliver their full promised benefits. However, the wave of digital innovation is not reversing course. This chapter introduces the digital technologies that look to have the greatest impact on the industry in the years ahead, where those impacts will be most prominent, and when those impacts will appear.

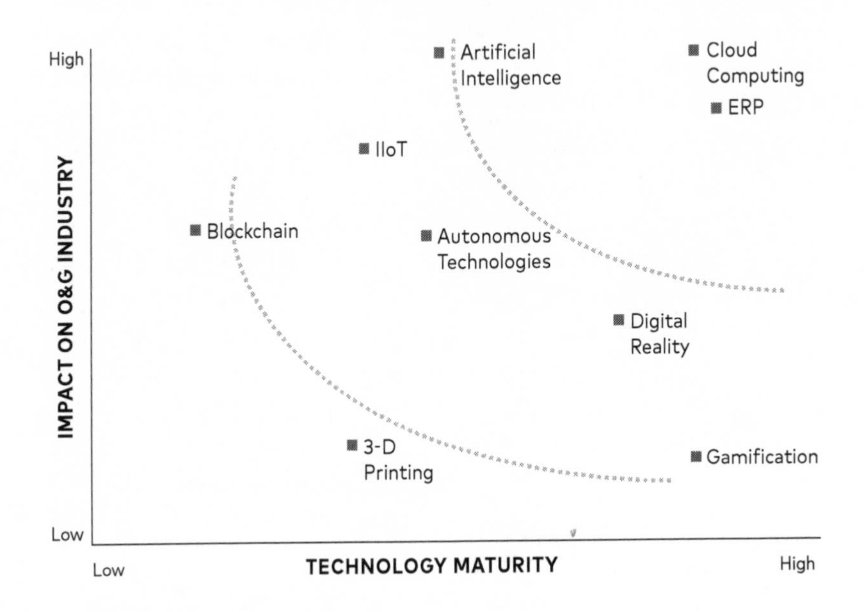

CLOUD COMPUTING

Maturity: high
Benefits: speed, flexibility, compute capacity
Uses: collaboration, data storage, rapid prototyping, platform for most other digital tools

Cloud computing is one of those exponential technologies that typify what I think of as "digital." As chips (for computing and memory) and bandwidth (for networks) have fallen to near zero in cost and dramatically improved in capacity, companies like Amazon, Microsoft, and Google have built huge data centers that house racks and racks of computers, acres of data-storage units, huge power supplies, lots of air-conditioning, and redundant global telecommunications access. This shareable computer horsepower is available to rent on a variety of commercial models (long-term or short-term usage) and for various needs (such as data storage, compute cycles, video streaming, analytics, and more).

Shared computer services offer a set of benefits that appeal to an increasingly large range of companies that previously relied on internal computer services. Cloud computing scales up and down quickly, which is really advantageous for things that go viral. The *Pokémon Go* craze resulted in 7 million new users in its first week alone. Only cloud computing could allow that service to run globally with high performance on that timeline.

Cloud computing works well for events such as the Olympics and other occasional sports spectacles like the Tour de France, which take place infrequently and whose audiences and demand for streaming and analytics are very large and unpredictable. The oil and gas industry also experiences occasional events, but of adverse natures, like spills and accidents, which attract large temporal and casual audiences. For all but the largest companies, cloud computing is a faster and more efficient means to satisfy the demand for emergency response centers, first responder and civilian coordination, and media distribution than having in-house standby facilities.

CLOUD COMPUTING IS having the same impact on industry that fossil fuels had on airplanes. In the decades before the famous Wright brothers flight, engineers had broadly figured out the intricacies of flight. The missing technology was a fuel of the right density to drive a sufficiently powerful motor to pull the aircraft through the air for liftoff. Once that fuel was mastered, airplane technology began rapidly advancing. Cloud computing similarly provides enough data and compute density to allow many other digital technologies, such as artificial intelligence, collaboration, and blockchain, to "take off."

Where fast-moving, lean, iterative, and innovative software developers require lots of computer horsepower on short notice for workspace, trials, and production systems, cloud computing is the solution. This enables a "fail fast" culture that accelerates innovation by eliminating the need for a large capital outlay for computers and long delivery timelines for in-house computer services. Many companies report that they can now run multiple large development projects in parallel because of the effectively unlimited computer capacity. Oil and gas companies often build bespoke solutions for their businesses, which could benefit from this kind of computing.

While cloud computing makes many companies concerned about security, cloud providers can actually offer superior services in areas such as physical security, redundancy, backup and recovery, telecommunications support, identity management, and performance analysis. It is simply more difficult for an internal IT organization with limited resources to keep up with all the changes happening in these various disciplines. Many security breaches are caused by overworked internal IT organizations unable to match the pace of patch-release from their software providers. Cloud providers often have the freshest knowledge of security developments because of their scale and reach.

Cloud computing enables the other digital innovations that follow in this book. Some newer solutions are inherently cloud-based, such as blockchain applications. The latest software solutions assume cloud-computing environments: developers are concluding that to reach the largest possible user base as quickly as possible, and to make the most amount of money, they are far better off to build for the cloud rather than for specific environments. For example, Adobe, responsible for the software that produces PDF versions of documents, has already almost entirely moved its product suite to the cloud.

Some of the best-known brands in the world have experienced the kind of growth rates that only an exponentially expanding technology could support, including Airbnb, Spotify, Facebook, Twitter, Apple TV, Netflix, Uber, and WhatsApp. Could these businesses have developed without cloud computing? Yes, but certainly not as quickly. Cloud computing has the potential to both improve the cost profile of existing oil and gas businesses and to enable new and transformed business models.

So far, there's no clear winner in the race to establish market dominance by a single large player—Oracle, Google, IBM Cloud, Microsoft

Azure, Amazon Web Services (AWS), and Alibaba Cloud. Amazon has a commanding market share lead of 45 percent, but a battle for market share is solidly underway.

Cloud services are bordering on free, with costs so low now that virtually anyone with even modest purchasing authority (that is, $2,000 per month) can contract for mission-critical business services. The low price point has been encouraging managers to go cloud, so it's common for large companies to have 200-plus cloud contracts for services that they didn't know about. Frankly, overnight hotel bills and taxi charges probably get more oversight than cloud service contracts.

Low prices for cloud computing do create business risk. For example, in Canada, it is illegal in some jurisdictions for certain kinds of data (such as private data on residents) to be stored on technology outside the province of residence. Cloud environments by the big players were not designed to meet these narrow requirements. They are designed for fast response times, redundancy, and scalability. A large company could easily find itself in violation of the law because, somewhere deep in the organization, someone contracted a cheap cloud service without doing enough research and due diligence.

Not all clouds are the same. Different kinds of compute loads tend to favor one cloud offering over the others, so companies will in all likelihood have a multi-cloud strategy. For some uses, a private cloud (only used by employees) may make more sense, while public clouds may be more valuable where the solution serves a large number of contractors, customers, and contingent workers, for example. In fact, 62 percent of cloud adopters find themselves with contracts from more than one cloud outfit. That makes a lot of sense if the goal is to use the best cloud for the job.

Existing web applications (accessible by a web browser) and mobile apps are migrating to the cloud first. Technically, the migration is swiftest for these kinds of solutions because they share important heritage elements (in design and in technologies) with solutions built for the cloud. Collaboration solutions that enable business integration with suppliers "outside the fence" are another attractive migration category. One of my favorites is the collaboration potential between upstream oil and gas companies and their suppliers.

A NEW INDUSTRY solution for oil and gas involves collaboration between services companies, asset owners, and workers. With supercomputers in the pockets of most workers, tablets in most vehicles, and connectivity to a network to access cloud solutions, these three participants in fieldwork can now operate in concert with each other. Services can be more tightly scheduled thanks to the greater visibility afforded by more real-time data and reports of work in progress.

The next important driver for the demand for cloud computing will be the addition of sensors on equipment. It is far faster to set these sensors up wirelessly via an Internet connection to the cloud, rather than a wired connection to an in-house data center. Companies will eventually turn to cloud computing just to cope with the massive amounts of data and analysis that results from lots of connected plants and equipment.

Cloud computing is probably a one-way trip. Once dependency on a cloud company becomes established, customers may be unable to object to arbitrary and unreasonable price increases. Changing strategies or cloud providers may be very challenging, particularly for large oil companies that measure their data holdings in multiples of petabytes, a volume of data that will not be quickly or easily shifted. Choose suppliers wisely.

ENTERPRISE RESOURCE PLANNING

Maturity: high
Benefits: reliability, data quality
Uses: core business processes, finance, supply chain, HR

One of the most substantial technology assets in use by many large oil and gas companies, as well as their suppliers, is an enterprise resource planning (ERP) system, and the most commonly used one is

SAP. A gigantic digital shift will take place in the next decade featuring the mass migration of oil and gas companies to the next generation of SAP.

Many large oil and gas companies are now wholly dependent on these big ERP systems to run their businesses. But these systems were designed for a world that no longer exists: their internal circuitry is optimized for scarce processing power and storage, two computing resources now in ridiculous abundance because of cloud computing. Their designs generally predate many of the things we take for granted in today's technology landscape: the Internet, mobile devices, cloud computing, app stores, big data, and analytics. Older versions of ERP systems are costly and difficult to maintain because of the huge range of databases and server technologies on which they operate.

Oil and gas companies operate the most complex ERP installations in the world, as measured by the volume and sophistication of customized code added to the basic software. Limitations in ERP capabilities compel oil and gas companies to customize the code by writing their own additions directly into the ERP software, restricting the ability to upgrade the ERP systems. Unique accounting rules in some countries like Nigeria have led to highly specialized ERP versions that are expensive to maintain and operate.

Despite the enormous range of capability in ERP systems, oil and gas companies have consistently felt the need to acquire and install countless other systems, leading to a huge portfolio of incompatible technologies all jostling for funds and attention.

SAP's response to the digital wave of change is very important to the oil and gas industry because so many companies run on it. Through its next-generation solution, SAP aims to reset its technology for the world of the future.

New Digital Features

With compute horsepower so cheap and abundant, thanks to chip costs that have fallen to basically zero, SAP's newest design, S4/HANA, calls for the entire system to reside in computer memory chips, rather than on disk drives. Disk drives have moving parts that fail occasionally; they're also slow and limited by the rotational speed of the disk. Chip speed is limited by the speed of electrons on a wire. By eliminating the need to move data to and from disk drives, SAP speeds up the system quite dramatically.

When I was studying computer science in the '80s, database design theory emphasized minimizing redundant and duplicate data. The reasoning at the time was that disk storage was costly, and a good database design minimized storage needs and the lease cost of a disk drive. This carried a price, of course. Data would be stored in tables (like an Excel spreadsheet), but important data that was closely related (e.g., an oil-well description and its monthly production) was stored in multiple tables. The solution was to join the tables and create a third table when you needed, thus keeping disk storage low and using computer processing to compensate. Sometimes this design can consume hours of computer time trying to marry up the tables to produce the needed report.

But in a world without computer limitations, why not eliminate all the myriad tables and have a simpler, flatter database design? It would be much faster to do reporting; indeed, some previously impossible queries would become standard.

Next-generation SAP has a flat database design. Users will be able to replace hundreds of slow and costly reporting procedures with streamlined reports and queries. Analytics that might normally take place separately can be done as the process executes, allowing different, better decision-making in the moment.

This new design streamlines the data tables, eliminating redundant data whose purpose was to help optimize disk usage. The savings in space creates capacity for all kinds of new data (from sensors, for example). Some back-office costs will disappear because of the pickup in speed and system responsiveness, and some processes will simply be reimagined so that they are daily rather than month-end.

SAP will become a web application, since it is much more cost effective to write systems for the cloud and a single-database technology that is accessed via web browsers. As a cloud solution provider, the publisher has only one type of database to support and only two or three hardware devices (Apple, Android, and Windows) to consider. Updates to the cloud version, or patches for security, distribute to everyone the instant a problem appears, or a new feature becomes available. No more lengthy wait times for an upgrade to happen. The risk of owning a stranded technology asset falls, along with data-center costs.

Past ERP designs predate smartphone technology and, for the most part, assumed that workers were desk-bound. Today and in the future, more and more work will be conducted via mobile devices. Indeed, a wave of digital innovation in oil and gas will come from equipping the workforce with smartphones. Mobile computing has its own challenges, including what happens when you're out of range from the network, the need to confirm the phone's owner, and the test of the phone's authority to update company data. Legacy ERP design did not contemplate the possibility that the end user could at one moment be online and the next offline, using a phone or tablet from anywhere on the planet.

What about all the new smart devices and sensors with built-in compute capability? ERP systems need to support these devices and incorporate them into work processes. The data these devices collect can finally be incorporated to provide much better analytics. I will not be surprised if my future car talks to an SAP system directly. Support for blockchain, a technology that will work well with the Internet of Things, will also be embedded in the system.

THERE IS A clear and strong business case to adopt digitally enhanced ERP technologies:

- Work will speed up.
- Data quality will improve.
- Costs of ownership (and support costs) will decline.
- The risk of a stranded but critical business system technology is all but eliminated.
- Many modernizations that have been elusive (mobility, or Internet of Things) are suddenly in reach.
- The power of in-the-moment analytics is finally achievable.
- Business will be nimbler—better able to embrace innovations faster and with less friction.
- Next-generation business models for oil and gas will be possible.

“Buy best of breed software and everyone is unhappy. Buy enterprise systems and everyone is still unhappy. But at least you can swap out best of breed. Enterprise is a one-way street.”

A CIO OF A LARGE OIL COMPANY

Most important is the power that comes from the ability to reconfigure business on the fly to address opportunities and disruptions. A shipment is delayed? What is the impact on the supply chain and what can we do right now to address it? A competitor has a problem? What can we do to quickly ramp up production to capture temporary volume?

Engineering a big shift from one generation ERP solution to the next is a fraught endeavor. There's a lot of complexity in the change, it takes time and commitment, and it alters every company agenda until it's done. As with any technology, the first to adopt faces the most challenges, including unexpected bugs and system issues. Business has also learned, from earlier waves of technological change, that such programs need executive commitment and robust change management to prepare the workforce for the new work practices to come.

Unlike other digital tools, which are compelling in their own right, SAP is imposing a time limit to tackle the migration by discontinuing support for its old versions by the middle of the next decade. Experience suggests there may not be enough skilled technologists able to undertake the mass migration if the oil and gas industry waits until the last minute. SAP is integral to the manufacturing sector, the public sector, banking, and so on—everyone faces the same migration deadline. Judging when to move (now when costs are low, resources are available, but risks are higher, or later when costs are higher but system-related issues have been resolved) will be tricky. Since big migrations can take some time, the largest oil and gas companies have already started to migrate.

ARTIFICIAL INTELLIGENCE

Maturity: moderate
Benefits: better decision-making
Uses: analysis of very large datasets, autonomy, modeling

I like to believe that intelligence is confined to our species. To me, intelligence is that special set of cognitive skills—such as visual perception, the ability to communicate with language, and decision-making—that distinguishes humans from other species (animals, fish, or insects), although researchers continue to conclude that some creatures, such as crows, rats, parrots, apes, monkeys, and dolphins, show signs of rudimentary intelligence. They note the presence of language skills, tool usage, and the intentional manipulation of the environment. Some clever macaques even seem to demonstrate a basic grasp of economic theory and the meaning of money.

Something that is artificially intelligent uses computers to do some or a significant portion of something a naturally intelligent species can do. It uses its computers to process a set of rules, uses sensors to detect the world, maintains historical data to provide reference to prior situations, and takes some action based on the outcomes of applying the data to the rules.

An autonomous car embodies quite a bit of artificial intelligence. It can detect and avoid obstacles, predict the likely path of other moving objects, and change its behavior according to conditions, such as slowing down in the rain or snow. Many robots in the future will feature similar kinds of navigational intelligence, learning, and adaptability.

Making something artificially intelligent is a job in itself. One approach involves showing a computer a thousand pictures of forks, so that eventually its visual sensors can look at an object and determine if it is a fork. Take a look at Quick, Draw! by Google, which encourages people to doodle online while an artificial intelligence engine interprets the drawings and teaches itself to recognize real-world objects from the doodles.

FIRMS IN THE EU have to comply with complex tax rules, and one of the most complex areas is the applicability and amount of value-added tax applied against goods being shipped across borders. There are thousands of individual tax rulings in this one area. Just finding and reading up on all the relevant rulings for a new case can take hundreds of hours. Companies want to know in advance their chances of getting a favorable ruling based on the specifics of the situation.

A tax firm has painstakingly loaded all of the thousands of accumulated case rulings into IBM's AI engine Watson. It can now listen to a tax lawyer ask a complex tax question, instantly search the thousands of cases to zero in on those relevant to the question, and can even provide the probability that a case before the courts will be successful based on the history of prior cases. A good tax lawyer is 70 percent accurate in identifying the relevant cases, given enough time. Watson is 90 percent accurate, in seconds.

The secret to AI success is to find those instances where AI can augment natural intelligence to accelerate analysis or improve the quality of human decision-making based on its ability to process huge volumes of data using a set of rules or heuristics.

AI in Oil and Gas

Conditions look to be ripe for AI in oil and gas. Much gray-haired experience has left the industry, along with their grasp of the implicit rules of how the world works and creating a need to capture and codify the rules. The environment has grown considerably more complex (more rules, more regulations), meaning there's a lot more data to process. The abundance of available resources has expanded supply, depressing prices and narrowing the margin for error.

New recruits to the industry, lacking experience, need to learn the rules quickly. Courts and regulators are unforgiving when the industry messes up, driving the stakes higher. Many oil and gas companies conclude that they use only a fraction of their available data for decision-making, and with limited human capacity, it's unlikely they will ever use that data without computer assistance.

One place where AI could help oil and gas is in dealing with contracts. Contracts and agreements in the average upstream oil and gas outfit number in the thousands. They're all different, often with unique clauses, triggers, and riders. They're hard to understand (they're written by lawyers, after all), and they can be in force for a very long time, outlasting their authors and original stewards. Knowing which are approaching key milestones is nigh on impossible. Just finding them all can be a challenge.

AI could cut this paper problem down to size in short order, in much the same fashion as IBM Watson has transformed tax law research in the European Union. Imagine using AI to quickly find out which contracts need attention and why and to recommend actions to management.

Artificial intelligence and machine learning could be brought to bear on resource data interpretation. As more oil and gas companies open up their data to cloud-based AI engines, those engines improve their performance, simply because AI can chew through more data more quickly than humans and learn faster. As with many other fields, AI applied to seismic data may change fundamentally how its interpretation is done. The future of geologic interpretation may well split into more routine, largely AI-led work and high-end, complex, creative human-led work.

Figuring out where to drill is a task tailor-made for AI assistance. Engineers spend upwards of 40 percent of their time assembling the data to set up a drilling program, according to estimates by Woodside Petroleum in Australia. They need data from prior drilling campaigns, actual costs, infrastructure locations, nearby well logs, seismic data, geologic interpretation, and so on. Not just any data, but the right data, and they have to highlight the specifics that require special attention. AI makes short work of this problem.

Dealing with field tickets is another good fit for AI. Oil companies still receive thousands of tickets from field companies for services rendered. Staff have to look at them, figure out which well they apply to, assign the right account codes, suspend them for investigation, even when they're neatly typed out on a spreadsheet and submitted as a PDF. Compliance reporting for commodities like water and emissions generates its own tsunami of documents. Some oil and gas companies actually employ more accountants and water-usage compliance teams than geologists.

AI could convert field tickets quickly and accurately into useful data, by using language processing to convert and interpret the text, identify and extract the right data, feed that data into the right systems, and make decisions to accept or dispute the charges.

Case Study #1: Watson

Some of Australia's west coast offshore gas facilities have been in production for decades, and its owners want the assets to be in production for decades more. The same ambition, a long and productive industrial life, is true for all similar offshore assets in Australia, as well as mature basins like the North Sea and newer resources like Canada's oil sands. Not only do these big assets handily outlast their designers, but they're now outlasting their original maintenance engineering staff, operations leaders, and logistics managers. In short, the complete original workforce.

But the oil and gas industry has long relied on the memory of its people to retrieve critical information about its assets, information beyond the kind of data easily found in modern systems. Answers to questions like "Why did we design it this way?" and "Have we encountered this problem before?" depend on the memories of workers. One of my clients once told me that the human workforce that manages complex assets involuntarily commit the assets to memory over time. It just happens. Human, or wetware, memory has worked reliably for decades, although it can break down: when turnover breaches 7 percent annually, corporate memory falters. A process that is highly dependent on people's memories is risky when the outcome is based on finding the right document or collection of documents and avoiding the wrong documents, under time pressure. The right people with the right memory might not be available, and memory can be faulty. When that documentation is important for in-the-moment operations, then risks will be higher.

Changes in the composition of the teams of people maintaining these assets reduces the effectiveness of the human-centric approach. There are more contractors and outsourced service providers. The average age of the oil industry worker, prior to the price drop in 2015, was fifty-five or older, and much of this gray gold has cashed in and left the industry to retire, taking their memories with them and leaving behind a much younger and less experienced team. Not only has turnover breached 7 percent, but for some companies, it's the 7 percent with the biggest impact on institutional memory.

Woodside Petroleum, the operator of the North West Shelf LNG project in Western Australia, is innovating to address this challenge. Like many similar facilities, the project has accumulated lots of studies over the years. Reports, diagrams, emails, meeting notes, investigations, spreadsheets, analysis—it's a daunting prospect to search all of its documents. The pile is growing all the time as the business matures and as the facility owners gives frequent thought to de-bottlenecking, capacity expansions, growth, cost reductions, quality improvements, and so on. The content is becoming more comprehensive, more complex, lengthier, and richer. Better tools, techniques, and technologies mean that studies can cover more ground. Instead of just one computation carried out by slide rule in 1960, modern computers can run millions of simulations, under different assumption sets, which are all captured in the studies.

In times of high oil prices and light regulation, facilities owners comfortably address these challenges by retaining large numbers of high-salaried employees, but at a cost. Woodside estimates that some 40 percent of engineering time is typically occupied with finding documents and reading those documents to discover what, if anything, can be useful. That's a lot of engineering cycles that could otherwise be devoted to more profitable activities, such as engineering. And when some piece of analysis can't be found, oil companies often find themselves purchasing the same analysis or data over and over simply because they presumed it was lost.

If there's one thing that computers are already very good at, it's sifting through mountains of documentation to quickly find things that match a set of criteria. Search engines like Google demonstrate their proficiency in finding web pages and documents based on just a few key words. Woodside has been feeding hundreds of engineering studies into the IBM Watson AI system, and Watson has been indexing the content to make it instantly retrievable. Engineers can now verbally ask Watson complex engineering questions, using technical jargon and terminology, and Watson returns every page of every document that matches or closely matches the question, ranked by best fit, within seconds.

The more precise the question, the narrower the results, and Watson helpfully presents the evidence to justify its reasoning. Employees teach the system over time which documents are most valuable in specific circumstances, so Watson gets smarter at interpreting questions and identifying answer sources that are more reliable. AI, in this instance, is akin to having the documented memories of every former and current engineer, every former and current contractor, every former and current specialist, and all of their accumulated expertise in one super-quick engineer who can do 80 percent of the job instantly, gets smarter with every question, never sleeps, and never takes shore leave.

In an industry with high levels of human expertise, there are many opportunities to augment that with artificial intelligence. For example, this same engineering application applies to many different facilities—oil sands plants, mines, refineries, petrochemical plants, gas plants, pipelines, and offshore platforms.

A second Woodside use case includes reservoir analysis, where finding and sifting through well logs, drilling records, land documents, and reservoir studies takes up considerable time by petroleum engineers.

Case Study #2: Digital Twin

Another branch of artificial intelligence is the complex modeling of the real world using mathematics and algorithms. A specific type of modeling that creates a virtual mathematical model of a plant (such as a liquified natural gas facility, a gas plant, a tank farm, or a refinery) is commonly called a digital twin. The idea of a digital twin is not novel to oil and gas. A visit to any geology/reservoir team will immerse you in the world of multidimensional seismic data, well logs, simulations, and resource models. In other words, a fully digital version of the underground resource. With the digital twin of the subsurface, the geologists can simulate how it might behave under different scenarios (such as various drilling and fracking programs). There's still the occasional news story about how some oil and gas company has built the latest edition of its subsurface digital twin using supercomputers, the biggest data centers, and most powerful processing power.

One of the truly great breakthroughs afforded by digital innovation in oil and gas is the ability to create a fully functioning digital twin of just about any asset or of an entire business. These newer versions include many layers of data that work together to provide a rich, fully integrated, and analytically deep software version of the asset or business, including:

- the engineering content (diagrams, specifications, and configurations) that describes the physical asset in digital terms for the engineering disciplines;
- the maintenance history (timing, procedures performed, parts installed, and installers) that provides insight into the ability of the asset to perform to its potential;
- the physical constraints of the various assets (operating capacities, throughputs, and pressures) that restrict how each asset physically behaves;
- the operating parameters of the assets (input energies, consumables, by-products, and emissions), which constrain the asset's performance;
- the financial aspects of the assets (fixed build cost and operating cost per unit) that yield the economics of the business; and
- the uncertain elements (customer demand, weather events, or supply disruption) that comprise the real-world conditions with which the business must cope.

The digital twin can have any number of these kinds of variables, whatever makes the most sense for the asset owner. For example, a digital twin of a wind farm would include the wind intensity over time, including day/night and seasons, and the demand for power. An oil refinery would include the crude slate with its variances of TAN values, sulfur content, and heavy metals; vessel-arrival rate; and market demand. A tank farm would include customer orders, supplier shipments, and blending opportunities. A digital twin of a mine would include the volume of material moved by its shovels, trucks, crews, and conveyors.

There's now enough technological horsepower available at relatively low cost that any business, including those as complex as oil and gas, power plants, and renewable energy farms, can build a fully functioning end-to-end digital twin and not just digital twins of its specific assets. The digital building blocks that make this possible are the fields of analytics, cloud computing, inexpensive sensors, and new programming platforms.

The quality and utility of a digital twin is very dependent on the builder of the physical asset. During construction, capital projects generate enormous quantities of data: who made the asset, the nature of the materials used, all the parts, the warranties, the costs, the expected performance profile. Multiple engineering disciplines contribute to the build—mechanical, electrical, structural—each with its own set of diagramming standards, taxonomies, and practices. Little thought is directed towards the future value of this data, so it's not unusual to find it spread out across many different proprietary systems.

For example, just five years ago, one large liquified natural gas (LNG) project in Australia (at the time one of the ten largest oil and gas developments on the planet) divided its capital project into four engineering contracts (the gas wells, the gas treatment plants, the large diameter transmission pipeline, and the LNG plant). At the conclusion of the build, the owner had a good view of how each asset was built because of the engineering data in those systems but then had to build a completely separate view of how the assets would work (the operating system). They had an ERP system that told them what the assets cost to maintain, run, and staff. The maintenance team had another system that tagged all the assets, issued work orders, and kept maintenance records. And finance then built models to try to predict the future economic performance of the assets under different scenarios.

Each of these systems held slightly different versions of the same data, incurring an extra cost and inhibiting analysis that worked across the systems, such as how to optimize the full business. The company then had to embark on a fresh initiative that integrated these systems, adding more layers of complexity and rigidity into the data environment.

I draw a distinction between the digital twin and the digital mimic. Many companies will stop short at delivering the full digital twin, and instead content themselves with something rather less. The mimic, like a parrot that imitates a human voice, cannot do all the things that the real thing can do. An industrial mimic might integrate engineering diagrams with the ERP system (a good thing, and very clever) but fall short of providing a way to visualize the asset and how it physically works under different conditions.

Some early adopters of digital twin technology in oil and gas are capturing benefits that are otherwise unavailable. The digital twin validates key assumptions and reveals which ones are no longer useful. Many businesses run on sets of operating assumptions—handed down from worker to worker—that may have been accurate when the business was first created but may no longer be valid.

The digital twin can reveal hidden-value opportunities, by improving use of key assets or avoiding certain high-cost ways of operating the assets. In times of capital constraint, boosting the turnover of existing assets may be the only available way to improve the business. Using a digital twin of their business, a tank farm operator discovered that their standard practices for moving product between tanks sometimes yielded a negative margin in a month, but that with better analysis of their portfolio of orders and shipments, they could convert that negative margin to positive.

A digital twin reveals bottlenecks and capacity constraints, which are often not plainly apparent in the financial results of the business. A good quality digital twin allows you to add, enhance, or take away assets to simulate its resulting behavior and inform capital choices.

AN ASSET OPERATOR used a digital twin to evaluate two capital options: which one would deliver the most benefits? They discovered that though their preferred choice had a low capital cost, it would fail at times of peak loading, which was undetectable using average or smoothed loading curves. Their alternative carried a higher capital cost but would not fail at peak times. By modeling out peak failure times and failure costs, the company concluded that the higher capital cost asset was the lower overall cost.

INDUSTRIAL INTERNET OF THINGS

Maturity: moderate-high
Benefits: visibility into asset performance
Uses: equipment monitoring, business optimization, decision-making, automation

One of the biggest spend categories in all of digital will be on sensor technologies that connect to the Internet, creating a future network of things. Physical devices, such as vehicles, equipment, valves, pumps, tools, and gear embedded with electronics, software, sensors, actuators, and connectivity will enable these objects to connect and exchange data. The value of connecting things to the Internet, and making them visible for business purposes, is only just beginning to make an impact in business. Digital technology companies describe these sensor-enabled devices as the Industrial Internet of Things (IIoT). Oil and gas companies have decades of experience with sensor-enabled things—it was called Supervisory Control and Data Acquisition (SCADA).

While the cost barrier for sensors, chips, analytics, and data communications is falling to zero, the benefits from adding sensors to things like pumps, equipment, tanks, and personal gear are starting to mount. These include:

- providing precise whereabouts of employees in the field for safety purposes;
- detecting with biometric readers when employees are suddenly prone;
- enabling higher-quality and lower-cost fieldwork using augmented reality;
- monitoring equipment vital signs like pressure, temperature, and vibration on tablets and phones;
- precise measuring of volumes, flows, and rates using cloud-enabled gauges;
- eliminating lost tools and rental equipment; and
- tracking inventory on the move to provide for flexible customer service.

There is no practical limit to where Internet-enabled devices could be deployed in industry. Suppliers to the industry will be highly motivated to add digital smarts to their wares as a means of competitive differentiation. All this data brings a tremendous value to the industry in areas like safety performance, energy optimization, precise asset utilization, sharper logistics, and so on.

The expected rapid pace of growth brings with it the need for managers to get in front of several important implications that this wave of sensors will impose on the industry. The immediate and most obvious implication is that these sensors are almost pointless without a network connection. To get the benefit of the data from the devices, the industry's various assets (such as offshore platforms, plants and refineries, wells, batteries, tanks, pipelines, and jetties) ideally need some measure of access to the Internet. Fortunately, sensors will not need robust, always-on, high-capacity connectivity; some digital devices will use only the barest minimum of network capacity to get the job done. New wireless technologies can blanket a large plant relatively inexpensively, without the need to dig trenches for cable.

AUSTRALIA'S VAST AND empty bush country lacks terrestrial telecommunications access. An innovative sensor company with a digital tank gauge has deployed satellite uplinks to squirt a tiny packet of tank volume data at certain times of the day when specific satellites are overhead.

As end points, Internet-connected devices will generate significant data volumes. Not all of this data will be of equal value: a sensor equipped with visual analytics (a kind of artificial intelligence) need not dispatch every second of video, but only that video, or even just a frame of video, that needs AI interpretation. Exactly how much computing and analytics that should be at the edge is an important design question, along with how important that end-point-generated data is.

Securing these device end-points from cyber threats is a high priority. Device-makers have not historically incorporated into their designs the kinds of cyber protections that are common to things like mobile phones and servers. Authentication, authorization, patch upgrades, and other processes typical of commercial IT are pretty foreign to the world of operations where such processes have not been necessary.

ONE OF AUSTRALIA'S natural gas companies experienced a failure of a SCADA computer that supervised a large battery of gas wells. It took some forty-eight hours to track down the engineer who was responsible for keeping the system running. As an operations system, it was designed to not fail, so there was no process in place for logging an incident, contacting a support team, escalating the failure, or patching the equipment. While help desk processes are standard fare for commercial IT systems, this situation is likely to unfold more regularly as sensor-enabled devices find themselves in the field.

Buyers of sensor-enabled devices need to be wary of unintended consequences. Unlike the broader digital industry with its open-source orthodoxy, technology innovation in oil and gas has often been closed. Proprietary technologies from the traditional suppliers to the industry may not integrate well with other technologies. Oil and gas technology has a long R&D cycle, measured in years to advance from innovation to widespread adoption. Suppliers need to recover their investments and, to do so, protect their innovations with patents and avoid open-source designs. The integrators of oil and gas technology (generally, the large equipment players) have built out their product families to try to satisfy the demand for equipment that works well together. In doing so, they have inadvertently (or intentionally) created walled gardens within which their technologies thrive, but other technologies do not. In comparison, open technologies are much more appealing for acquirers because of the ease of integration into the buyer's context.

Compare this situation with defense and military technology companies. The design of technology for military field use must be kept open and fluid to allow for innovation and advancement, such as smart bombs, improved radar, better protective clothing, field communications, and so on. Military buyers would not invest in a billion-dollar fighter-aircraft purchase if those jets would be frozen in time and unable to take advantage of the latest weapons technologies. Military-grade equipment must be plug and play. Indeed, a global supermajor retained a defense contractor to oversee its next oil infrastructure project to bring that kind of approach to the design.

Buyers should also be mindful of data standards for sensors and devices. According to the IEA, the International Electrotechnical Commission recently revealed that there are more than forty different bodies that have set standards applicable to cybersecurity, amounting to some 650 different standards. This confusing state creates barriers to achieving interoperability. Since it takes time and costs money to investigate a standard and figure out how to comply with it, and with so many possible standards, digital innovators are understandably reluctant to engage. This has not been the case with equipment companies—industry has fewer standards bodies, and compliance with those standards is usually a mandatory requirement in a procurement situation.

A final consideration for buyers of sensor-enabled devices is that companion technologies, like artificial intelligence and machine learning, become mandatory. Internet-enabled things will generate enormous quantities of data and tools like Excel are not robust enough to process these volumes. A single drone flight will generate much more data than can be comfortably analyzed by even a big desktop computer. As device investments expand, and users progress from monitoring devices to analyzing the data from devices, companies will need to invest in analytics.

Swapping out dumb iron with smart iron will take time because of the management of change (MoC) processes in place at virtually all oil and gas installations. MoC provides for the safe execution of changes to oil and gas infrastructure and includes such steps as the preparation of technical feasibility, detailed change specifications, quality assurance reviews, cost and benefit analysis, careful implementation planning, notifications, safety reviews, inspections, and so on. Without an MoC process, changes could create unsafe conditions.

Over time, equipment purchases should trend towards equipment that is more digital, as owners aim towards new facility designs that are more automated, more self-managing, and more reliable. Equipment makers that do not feature some digital capabilities with their equipment will eventually fall out of the running, except perhaps in replacement procurements where like-for-like is the only plausible solution (e.g., replacing a dumb valve with another dumb valve).

AUTONOMOUS TECHNOLOGIES

Maturity: low for devices, high for processes
Benefits: reliability, cost, productivity, safety, access
Uses: drones, vehicles, submersibles, robots

Of all the digital technologies that ought to encounter a receptive market in the oil and gas sector, autonomous equipment should be greeted warmly. Moreover, it is in the best interests of many oil and gas basins to accelerate autonomy solutions. These days, many different kinds of devices meet the definition of an autonomous or nearly autonomous vehicle, including robots, aerial vehicles, submersibles, and drones, which are unmanned remote-controlled devices. A few universities, such as Wichita State University, are working very closely with industry on dozens of equipment experiments showcasing how autonomous technologies are advancing.

Digital advances are making it possible to safely operate many different kinds of equipment autonomously that formerly would have required human controllers or drivers. These advances require complex mathematics, learning systems, sensors, data networks, cameras, robotics, and digital controllers, which have fallen in price and expanded in capability to bring autonomous control within grasp of most manufacturers. Working sites with autonomous onshore and offshore drill rigs, rail locomotives, aerial drones, submersibles, and container yards are in place, and field trials are underway in delivery vehicles, cars, forklifts, airplanes, helicopters, construction equipment, cranes, and ships.

Autonomous equipment should be a sensible choice when the work required is some potent combination of extra dangerous, so elaborate protections are needed to safeguard workers; costly and repetitive, such as driving; or extra costly, perhaps by virtue of its location, such as in far Northern oil mines, or by virtue of the scarcity of the skills needed, as with pilots.

Some oil and gas basins, notably in Canada, the U.S., the UK, and Australia, have other intrinsic features that make them conducive to autonomous kit. These include a vigorous and technically capable supply chain, widespread network coverage, generally high-cost labor, demanding safety regulations, a large installed base of infrastructure, and low technology transfer costs.

THE OIL AND GAS sector is very well suited to adopting autonomous gear.

- The work can be dangerous—fumes and vapors from oil and gas can asphyxiate or ignite.
- The work location is frequently remote and harsh, driving up the cost of housing human operators.
- Well-trained human capital is now scarce because of the recent downsizing in the sector.
- The physical assets (plants, pipelines) are long-life assets that require steady and often routine attention, making payback less risky.
- The work has lots of routine elements (heavy haulers trundling around mines, or operators driving around to inspect facilities).
- Environments where autonomous kit can be deployed are not shared with public use infrastructure (which takes away risk).

Case Study #3: De-manning Gas Fields

Australia has an unusually impressive track record in innovating in the area of autonomous equipment in its resources sector. The new gas fields in Queensland are using drones to fly inspection runs over the assets (mostly gas wells). The wells are relatively low productivity, heavily regulated, numerous, and remote.

The old model was to assign a field operator to a handful of wells that would be visited on some scheduled basis. The ratio of operator-to-well was low, because of the extensive drive times to get to the wells. For at least one owner, well operators were on the same kind of shift as an offshore oil worker (two weeks on, two weeks off). Workers needed to drive enormous distances (creating a serious safety risk) and visited wells that needed no attention at all (leading to wasted effort).

Initial drone trials led to the realization that aerial technology could do the job using industrial-strength drone technology from serious players. The drones needed to fly at night so as not to disturb pastoral lands and grazing animals. They needed to fly quite high (at several hundred feet), out of sight, in a range of weather conditions, which required a strong drone engine and high-end aeronautic controls.

The drones also needed to carry a hefty payload—high-resolution digital cameras, emissions sensors, LIDAR (Light Detection and Ranging), moisture sensors, and so on. They took before and after photos to see how much vegetation has grown, if the well has been flooded, or if it's been damaged by kangaroos. The drones' data was plugged into work systems that created the next day's work roster, including what wells needed to be visited and why, what parts to load on the operator's truck, even the order in which to visit the wells based on landowner access permits and most efficient routing.

The payoff has been significant: a single drone can fly over and inspect more than 150 wells each flight, where an operator might visit just ten per day. De-manning the gas field delivers all the same HR benefits as autonomous haulers, with better safety outcomes because there's less driving. Of course, drone-controlling pilots are expensive but there are far fewer needed than drivers, and, because of their training in flight operations, they are inherently safety conscious. Unmanned aerial vehicles (UAVs) could have a big impact on the huge and widespread infrastructure that is typical in the Canadian and U.S. oil and gas fields (such as the Western Canadian Sedimentary Basin and the newer unconventional fields in the Horn River and the Montney).

Case Study #4: Autonomous Mining

The big iron ore mines in far-off Western Australia are considered world leaders in developing the autonomous mine truck. These heavy haulers work in a well-defined and controlled space (an open-cut mine) where there's no one about but the miners. The cost of a heavy hauler driver is very significant—multiple shifts of crews keep the mine operating around the clock, they stay in company-sponsored fly-in/fly-out camps and travel from all over Australia. The mines pay for the travel, pay for the camps, and pay extra for the social costs of supporting workers who are separated from families and communities for long stretches of time. Canada's oil sands mines are also adopting robotic mining equipment for the same cost-saving reasons.

The autonomous haulers deliver fleet-based learning—each hauler has an identical computer system that controls the vehicle, but they teach each other, creating a single virtual learning machine. Driving experience is captured and shared across all the haulers in real time, making the system constantly smarter and safer. The remote-control room that supervises the haulers can be anywhere in the world where the mining company can find workers, improving turnover and reducing absenteeism. Shift changes take place with no reduction in productivity because the trucks don't actually stop trucking.

Autonomous equipment, unless it is programmed to do so, will not vary from its assigned path. As a result, mine roads need more grading and maintenance to repair ruts caused by the repeated truck movements in exactly the same place. On the plus side, an autonomous truck can optimize its operations far more precisely than a human can, accounting for road grade, load, fuel cost and consumption, braking capacity, and so on. Precision hauling yields more accurate business performance.

TRIALS INVOLVING AUTONOMOUS equipment are plentiful in oil and gas, such as:

- the autonomous drill rig that cuts the human-operator head count from dozens to a handful,
- the robot trials for carrying out tank inspections and welding jobs during turnarounds, and
- the UAVs that carry out flare stack inspections (a particularly hazardous and nasty job).

There are many other trials of drones underway in oil and gas. Inspection drones operate inside fuel tanks during turnarounds to look for faults and cracks. Submersible drones execute seafloor operations in the North Sea and the Gulf of Mexico. Aerial drones conduct safety flights around the perimeter of refineries to monitor for breaches, inspect flare stacks and towers, and provide eyes-on views of plants. Drones monitor large infrastructure sites from the air to gather accurate readings of construction progress.

3-D PRINTING

Maturity: low
Benefits: speed, carbon avoidance
Uses: jigs and tools, prototyping, parts and spares

Additive manufacturing, also known as three-dimensional or 3-D printing, is a production technique invented this century. A 3-D printer interprets a digital design and sprays a thin layer of material (such as plastic, resin, composite, cement, or steel powder) from a nozzle to build up an object of that material layer by layer.

The designs for 3-D printed objects can be stored in the cloud and then transmitted directly to a 3-D printer with access to the Internet. Designs become collaborations between a customer, the designer, the engineering disciplines, materials specialists, the printer, the packaging team, logistics, the parts installers, field service, and maintenance, all working together on a single, shared design.

3-D printing technology shows the same kind of exponential performance gains as many other digital things, which is not surprising since the technology behind 3-D printing is the usual suspect (namely, computer chips). In 2008, the slowest 3-D printer cost $18,000. By 2017, the slowest 3-D printer cost $400 and was 100 times faster than the 2008 model.

Some industries have concluded that much value is at stake in 3-D printing. First, 3-D printing produces less waste. Only what is required is printed, unlike a traditional process that typically starts with a solid block of material and uses tools (like lathes and drills) to sculpt away excess material to reveal the final part. Less waste means less carbon invested in producing material that is ultimately scrap. Second, if designed well, 3-D printed parts can be lighter in weight, lowering shipping costs. Third, instead of producing multiple parts that have to be fitted together, 3-D printing can often print a single complex part, saving manufacturing time and eliminating more carbon emissions from shipping multiple parts. Fourth, instead of waiting for a part to be produced and shipped from a factory, parts can be printed on-site, yielding more time, shipping, and carbon savings.

Setup time is eliminated. A 3-D printer simply downloads a new design and starts the printing process. Packaging for shipping could be reduced. Jigs and guides for traditional machine parts are themselves 3-D printed to save time and cost. Finally, inventories can be lower as items are printed as needed.

MANY INDUSTRIES ARE experimenting with 3-D printing technology.

- A major tire company has designed a new tire inspired by honeycombs, which performs like a traditional tire but with a fraction of the material and without a need for air.
- Custom-fit footwear, particularly running shoes, has the potential to unleash an entirely new industry of bespoke designs, colors, and shapes to meet a customer's needs.
- Toys made of plastic and composites can be printed at home.
- Artists have discovered that 3-D printing allows them to experiment with new shapes and designs that cannot be produced using conventional techniques.
- Like footwear, next-generation eyeglasses made of composites will be 3-D-printed to match an exact head shape and design concept.
- Really large printers use building materials, like cement, to print buildings, bridges, and supports.
- Elaborate desserts are 3-D printed and presently available for purchase in drive-thrus in some Asian markets.
- The U.S. Navy has 3-D printers onboard its ships because 3-D printing a part is faster than having it shipped in. Of note, the Navy is running trials to print new submersibles for specific missions in a few days, rather than the six-month lead time from traditional supply methods.
- Aerospace companies print non-critical replacement parts for aircraft, such as door handles and latches, avoiding complete door replacement.
- Marines are using 3-D printing on the battlefield to make replacement parts like hinges and bumpers, and even equipment like snowshoes.

The Business Impact on Oil and Gas

It's not immediately obvious where 3-D printing might impact oil and gas. This industry begins with drilling or mining for the resource and continuously processes crude oil into useful products like gasoline and jet fuel. Product distribution is a supply-chain function with ships, pipelines, tanks, and trucks. So far, oil and gas trade shows don't give over much stall space to showcase additive manufacturing, based on my observations. The industry is used to having its kit made in large metal bending and hammering shops with heavy-duty forging and welding.

Hard-to-replace parts, critical spares, and rogue parts that chronically leak or fail could potentially be printed rather than machined, which would save some time and cost by avoiding unplanned downtime. Complicated equipment with many parts could be simplified. GE is already printing some petroleum kit, including valves and turbine parts, because printing allows for more intricate parts produced faster and to tighter tolerances.

Taking a page from the Navy, offshore oil platforms and rigs could use a 3-D printer onboard to produce some of those minor items while gaining experience in designing and printing larger more composite items.

3-D Printing and Demand for Petroleum

The far bigger and more immediate impact of 3-D printing will likely be felt in the demand for petroleum. A tremendous amount of today's demand for oil (about 50 percent) is for transportation markets—planes, trains, and automobiles. Much of that transportation is to move pieces and parts along supply lines.

Consider the amount of oil in a typical running shoe. The design could be from California, but the upper might be made from cotton that is sourced from Pakistan, which is woven and dyed in India and cut in Vietnam. Lowers may be cast in factories in Eastern Europe, but stitched to the uppers in Chinese factories. Eyelets are punched out in Taiwan, laces might come from Colombia, with plastic eyes attached in China. The cardboard box could be from the U.S., ink and labels might have been printed in Germany, final packaging is likely completed in southern China, whereupon the shoes are finally shipped to U.S. markets.

This might be the most cost-effective way to produce the shoe a market wants, but it produces a lot of carbon emissions. A shoe company once tried to calculate how much carbon is in a running shoe, and they were shocked to find out that it's measured in pounds per pair.

Many manufacturers are searching for ways to reduce their carbon footprint and, for some, using additive manufacturing will be a key tool in their arsenal, because it shifts both their "inside the fence" carbon footprint and their "outside the fence" carbon pollution from their supply chains. The result will be a reduction in demand for oil as supply chains ship printing materials directly to where the final product is printed and consumed, rather than shipping various incarnations from country to country.

The easiest course of action for an oil and gas company is to wait for 3-D printing to mature. The market will ultimately decide whether additive manufacturing has a role to play in the sector. However, oil and gas industry analysts, such as the IEA, forecast that 3-D printing will be instrumental in managing the carbon footprint of all industry, including oil and gas. The benefits are simply too big to ignore.

DIGITAL REALITY

Maturity: medium
Benefits: improved decision-making
Uses: collaboration, training, data usability

I first became aware of augmented reality by watching American football on television. On my screen, I could see information that was clearly not available to someone in the stands, such as the distance to the goal line, the number of yards remaining in a down, and a player's latest stats. My reality of the football game had been augmented with information on the TV screen.

Augmented reality (AR) is poised to become a big thing. A clever augmented reality game, *Pokémon Go*, captured the world's attention with 50 million downloads in its first week. Developments in the entertainment sector are generally not of great interest to oil and gas. In this case, however, these reality "toys" can be used to deliver real-world savings to oil and gas operations.

AR merges digital data and a real-world scene on a single screen. That screen could be the TV (as with my football example) or a windshield, a smartphone screen, eyeglass lenses, or a headset. The combination of the digital data and real-world scene creates a new way for users to interact with the real world. *Pokémon Go* presented imaginary figures to capture in the real world of parks, fountains, buildings, and monuments, all on a smartphone. A windshield might show your driving speed and the route to your destination. A fighter pilot's headset will show altitude, other aircraft, fuel levels, and so forth. At most oil and gas trade shows that I attend, someone will be demonstrating an augmented reality tool, usually with Microsoft HoloLens.

Virtual reality (VR) creates a visually immersive and consuming environment wherein the real world is completely replaced by a computer-generated rendering. VR requires a headset that blocks the user's perception of the real world. Visit the neighborhood consumer electronics shop and there will be an expanding assortment of such headsets that work with smartphones and tablets. For the most part, the content available tends to be short movies and games intended to educate the market on the technology's potential.

Mixed reality (MR) builds on AR and VR technology, combining a virtual world and the real world where both digital and physical objects—and their data—can coexist and interact with one another. In a manufacturing setting, a maintenance worker with a screen built into their hard hat could stand in front of a piece of Internet-enabled equipment and query it directly to determine its operating state, in addition to seeing relevant diagrams for its maintenance and repair onscreen.

Collectively, AR, VR, and MR are referred to as digital reality (DR).

Where DR Creates Value

DR can create a visual and engaging bridge between people, assets, places, and information where current business practices or business structure make it costly or difficult to bring these four things together. For example, oil and gas assets are often in inhospitable places (such as the Arctic), inaccessible places underground, or both (such as the seabed). Sending a worker to perform visual observations and maintenance is costly and, at times, dangerous, if not impossible. In cases where sensors and drones are unable to fully displace the need for a human worker, DR can help them execute their work.

In other instances, oil and gas assets or the assets' future location may not yet exist in the real world but do exist in the form of maps, drawings, and 3-D renderings. Engineers complete these drawings and look for issues related to safety, construction, and integration in the design, but flat diagrams fall short of enabling full understanding. This results in design errors, critical bottlenecks being missed, infrastructure being forgotten, and so on. DR takes this process to the next level by enabling engineers to virtually walk through, around, and inside their designs to detect quality issues, safety concerns, and fabrication improvements.

Finally, assets need maintenance, turnarounds, and shutdowns in a constantly changing world of workers, practices, compliance needs, and regulations. Hands-on training is the most effective way for humans to learn (by doing), but such education is typically unavailable in an at-scale energized asset or an offshore asset. Traditional training sessions are expensive and not fully effective in addressing this gap. Using DR training modules with a realistic simulation of the asset would allow turnaround teams to become virtually aware of the work site, key work practices, safety procedures, and team coordination.

Examples of DR

The automotive industry is a vigorous user of DR. First, AR and VR are used to design the car from the driver's point of view—how the car will look and feel to a driver. AR is used to design the factory where the car will be built—auto plants are huge, sprawling facilities with lots of cranes, robots, and equipment that are hard to embrace with 2-D diagrams and charts. Finally, AR and VR are used to build the car, showing how the parts fit together and enabling collaboration between design and manufacturing, between the automaker and its supply chain.

While oil and gas is only just beginning to explore DR, there are already some early successes. Some offshore oil and gas basins are in water too deep for human divers, yet the assets on the seafloor still require maintenance and services. This is a job for robots and autonomous submersibles, but DR can enable a fully immersive experience for off-site workers who are directing these unmanned robots. Shell has spent many years developing SensaBot, a robot equipped with sensing technology, cameras, and arms to carry out maintenance, surveillance, and rescue operations undersea.

Oil and gas, like power and utilities, or rail and mining, features assets that are widely dispersed. When those assets need inspection or servicing, current practice is to dispatch a team with all the requisite skills and experience for the work they are likely to undertake. However, through DR, an on-site worker can interact with a remote expert, share the same view of the work site, and have instructions from the remote worker appear on the heads-up display, headset, or even a tablet. Together, they can engage in joint problem-solving, work-step review, and safety briefings. The remote expert does not need to travel to the worker's location to be fully immersed in the situation. Fieldbit is an early solution in this area.

A third example is in using DR to create a virtual version of the asset, by bringing together 3-D drawings and asset data in a single virtual environment. Workers can experience the assets before they are built or, in the case of existing facilities, before mobilizing to the asset. Through DR, teams can carry out virtual de-bottlenecking and safety inspections, such as those being undertaken by the Linde Group, which enable assessors to proactively identify issues and address them at the blueprint stage. These same DR environments provide a training environment for staff before they arrive on site. Using virtual worlds, staff can be trained on maintenance routines and many more scenarios and incidents. Human nature, communication gaps, work decisions made with imperfect information, and short reaction time mean that staff rarely get everything right the first time.

Next, DR and geographic data (or geocodes) together can highlight where assets may be buried or hidden from view. Pipelines and cables can appear onscreen in place, showing workers exactly where drilling, excavation, and tunneling work can be done safely. In case of a fire, workers wearing DR headsets could be directed out of the danger area through other visual cues, even if their path is obscured by smoke, using digital technologies like PoindextAR.

The most exciting developments involve the ability of AR to interact with devices equipped with sensors (the Internet of Things) to deliver highly context-specific information to workers at precisely the right time. For example, a sensor could deliver a notification as a warning pop-up in the worker's view when equipment is energized and can highlight which parts of the equipment can be safely serviced. On-demand training and reference information can be triggered when workers access machinery that has been recently replaced (e.g., a worker is forbidden access until they watch a quick instructional video).

Training and reference information could be delivered via a headset or a smartphone or tablet. FuelFX allows users to hold up a tablet or smartphone device, aim the camera at an object, and see animation and details related to the object onscreen. Studies from the aviation industry show that workers using an AR headset displaying a maintenance manual are up to 30 percent faster and 90 percent more accurate than peers using only PDF instructions.

Moving Forward with DR

To surface plausible candidates for DR application, a useful starting point is current work processes that are costly or inefficient because of inaccessibility, invisibility, and information challenges. These challenges can be almost anywhere; as with the automotive example, DR may be able to improve design, build, operate, and maintain work processes, where assets are awkwardly located, hidden from view, or merely large in scale. Typical constraints noted by early users of DR include the availability of high-quality information to be displayed on screens, the sophistication of sensors and their ability to work together with DR devices, and the capability of the DR hardware to cope with the rigors of the oil and gas operating environments—cold, wet, and dark.

GAMIFICATION

Maturity: high
Benefits: accelerates adoption, personalization of technology, youthful appeal
Uses: safety, compliance, training, loyalty

Gamification means bringing insights and techniques that encourage competition, addiction, and execution excellence from the video game industry to other areas of commercial activity. Online games have been around for my entire business lifetime of thirty years, but only in recent years have video games become a big industry.

My first video game was mildly addictive. I worked for a summer at McGill University in the '80s, and we played a computer game that involved shooting at asteroids. Games were pretty rudimentary—there were no screens or game controllers. A dot matrix printer pounded out images of a slowly moving target composed of asterisks, slashes, and dashes, and the player typed in their firing sequence using a keyboard. Students tried to collect those-end-of-semester user accounts with a little leftover computer time just to play the games. It was all so hi-tech at the time, but, looking back . . . not so much.

Today, entertainment games feature a set of elements to create addictive behavior—point-scoring, rewards, personalization, sensory immersion, competition with others individually or in teams, deep realism, and various rules of play—and fall into categories such as simulations, adventure, strategy, puzzles, action, combat, role-playing, or training.

Modern examples of gamification in industry have focused on using games to encourage customers to engage with a product or a service. Retailers have been particularly aggressive in using games to lure customers. A fine example is fashion designer Norma Kamali's 3-D movie and online shopping experience, where Kamali challenges customers to find specific objects in the movie. Bonobos, the global men's retail chain, lures shoppers to spot specific items from its clothing line hidden on its website in exchange for shopping credits.

The foundational elements to bring gamification to oil and gas are falling into place. The engineering profession now uses software tools exclusively to design oil and gas infrastructure. That same data creates the rich environment necessary to render a purely online and immersive playground for games. Operational systems (SCADA) generate the kinds of real-time data to make environments even more real. Cloud computing provides ample computer time at low variable cost to house a game and play it across multiple companies, countries, and sites. The device world of smartphones, tablets, and eventually hands-free devices, like HoloLens, has put a game controller into everyone's hands. These devices can provide a solid gaming experience—with sound, two-way communications, video, and haptic response. Rapid developments in artificial intelligence, neural computing, and big data enable games for complex business systems like oil and gas.

A particularly strong opportunity for gamification is in capital projects. Have you ever seen the project plan for a large oil and gas project? Building a big facility is a collaborative effort involving a dozen key disciplines and often multiple firms, each of which brings to bear its particular set of assumptions and orthodoxies about how the world works. After many meetings and workshops, engineers emerge with a complex Gantt chart incorporating thousands of lines, written in engineering shorthand, and programmed into the industry's planning tool of choice, Primavera P6, a project-planning tool from Oracle. P6 is the dominant software tool for such large-scale project planning.

Suffice to say, these plans inevitably incorporate their share of bugs and mistakes that are in plain sight but hidden from easy detection. Every engineer has their capital project story of how a crane arrives to carry out a lift, but the items to be lifted haven't arrived, or the crane needs to navigate an impossibly tight space to reach the lift site, or the beams to be lifted do not have foundations in place.

What if the P6 plan could be programmed into a computer-simulation game engine that shows the work steps in a kind of slow-motion video? That way, the video could help reveal the inconsistencies hidden in the plan. Teams of engineers can "play" the capital game over and over, tweaking it to remove cost, speed up execution, and improve throughput of machines and crews.

Remember that scene from *Top Gun*, where Maverick and Goose replay a dogfight in the classroom? Top gun flight schools carry out a forensic review of missions to learn what worked, what didn't work, and how the team can improve. A second useful application of gamification in capital projects is in forensics. Often at the end of a large capital project, the engineering contractor presents the owner with the list of reimbursable overages that occurred during the project. Many are legitimate, but some should be turned back because the contractor did not begin the work with an efficient and effective plan. By loading actual job activity into the same game engine, the resulting game shows where the contractor's execution was inefficient and ineffective. Owners would be less willing to simply pay for contract overages, and contractors would learn to sharpen their plans.

Gamification techniques also apply to worker training. Employee health and safety training in oil and gas invariably includes basic awareness of the many hazards that exist on the jobsite. But translating the abstractness of pictures and PowerPoint into real understanding of the hazards and putting that training into tangible action on the job is still daunting. Incidents still happen too frequently. Using augmented reality, drone photography of overhead scenes, and game puzzles like "spot the unsafe condition," employees can "up their game" before stepping foot on-site.

These days, incident-recording using smartphone apps and cameras creates a never-ending flow of game content for safety training to keep these games relevant and updated. The oil and gas industry does not compete on safety but typically share incident data to help the overall industry improve.

Commercial optimization is a third possible use of gaming. The digital twin of an asset or a business provides most of the data needed to build a game version, which would provide employees with a safe virtual environment to learn through play the nuances of how the assets are likely to behave in real-world conditions. A fully functioning digital twin of an asset or a business includes many layers of data that work together to provide a rich, fully integrated and playable software version of the asset.

The game for an oil refinery would need to include the crude slate with its variances of TAN values, sulfur content, and heavy metals, along with the refinery complexity and its logistics setting. A tank farm would include customer orders, supplier shipments and blending opportunities, tank volumes, throughput, and pump capacity. A trading operation would include product demand, refining output, market pricing, margins, pipeline availabilities, crude choices, and trading strategies. Using the digital twin of the business as the basis for a strategy game could help companies train new managers, sharpen operating chops, set more meaningful performance targets, and identify opportunities to improve the business.

BLOCKCHAIN

Maturity: low
Benefits: origin-tracking, contract settlement, identity
Uses: trust-based processes, contracts, collaboration, disputes

One of the most transformative technologies yet to make a big impact on oil and gas is called blockchain, or distributed ledger technology (DLT). Media stories highlight the rise of new blockchain business models and cryptocurrencies, with space-age names like Bitcoin and Ethereum. However, there is so much more to blockchain than currency, and this technology will eventually have a dramatic impact on aspects of oil and gas.

What exactly is blockchain? Imagine a set of records, like the rows of a spreadsheet, which looks like a ledger book. My bank account is an example of a ledger—it has an opening balance, with spend and receipt transactions as rows sorted by date. Instead of this ledger being held in just one place (such as my bank), a blockchain ledger is stored on many computers (hence the word "distributed" in DLT). To replace the security of a bank, which holds a single copy of my bank records in their vault or computer system, a distributed ledger is encrypted, the records are linked to one another (hence the word "chain") so that they cannot be tampered with, and the ledger is replicated on many computers.

BITCOIN IS ONE way to implement distributed ledger or blockchain technology. Transactions that transfer Bitcoin from one account to another make up the ledger. Every ten minutes, 2,000 such records are grouped together into a block and are fed into an encryption algorithm called SHA 256, along with a random number (called the nonce). SHA 256 returns a value, called the hash. Bitcoin miners race to guess the value of the nonce that yields the hash with the most leading zeros. It's hard to find the right nonce, but easy to validate in that everyone can replicate the SHA 256 calculation once they know the nonce. The nonce is then attached to the block of 2,000, and the process starts over for the next block. The reason blockchain is so secure is that changing just one letter or number in the block of 2,000 returns a completely different hash, meaning every miner can immediately detect an attempt to replace a correct transaction with a fraudulent one.

Entries to this distributed ledger are added by companies (called miners) that maintain the records. Further, entries can only be added in blocks of transactions (hence "block") and only if the majority of the miners agree that the transactions are proper. This makes blockchain very hard to tamper with, because so many parties have to agree, and very hard to corrupt, because the transactions are linked to each other. Miners are themselves paid with Bitcoin for maintaining the ledger. With the power of global telecoms networks and cloud computing, market participants (such as buyers, sellers, and miners) can create highly trustworthy shared information.

Since a ledger is just data, it begs the question of what kind of data could be stored on such a distributed ledger. How about a registry of loyalty points or tokens (representing a store of value) which could be awarded to customers for repeat purchases of fuel? We are only just beginning to understand the full range of possibilities, but I like the mnemonic ATOMIC coined by William Mougayar to describe blockchain potential (see sidebar).

WILLIAM MOUGAYAR IS a Toronto-based blockchain researcher and writer. His work concludes that six key social constructs, individually or together, provide the basis for the potential application of distributed ledgers:

A—assets—any physical or intellectual asset that can be characterized with descriptive attributes. Diamonds, cars, music, and movies are examples.

T—trust—any relationship between multiple parties that is low in trust.

O—ownership—any relationship between parties or things involving ownership. Assets usually have an owner.

M—money—any relationship involving the exchange of money, such as trading and settlement.

I—identity—when the identity of parties or things is important. In banking, these are called "know your customer" rules.

C—contract—when the relationship between parties, assets, and money is codified.

Blockchain is a digital solution to one of modern society's basic problems—that we can't always trust everyone or everything, or that we shouldn't.

Many business transactions are based on the notion that we trust each other enough to do business, but not so much that we dispense with record-keeping. That's why buyers issue purchase orders, shippers issue packing lists, sellers submit invoices, and banks provide deposit slips, all of which are supported by agreements, contract terms, and numbering schemes that enforce tracking, delivery, and payment. These documents are representations of one party's ledger.

Privacy laws and confidential information rules are all linked to this question of trust. The whole notion of a privacy law is predicated on the lack of trust between parties, and the risk that one or the other might reveal some confidential information when it suits them. Will I hand over my health records to a hospital if I think they might hand over my state of health to my insurance company?

The reason blockchain has come into its own is its ability to dramatically lower cost and reduce misunderstandings, disputes, and fraud. A significant amount of overhead in a standard company is devoted to the various mechanisms it puts in place to assure trust with counter parties and to minimize the costs of eventual misunderstandings, disputes, and frauds. This includes writing and tracking all the outside contracts it has in place, the reporting it carries out for compliance, and the monitoring it self-imposes. And all the associated paperwork.

Not only is the volume of such tracking and compliance accelerating, the people hired for this job are higher-skill and higher-cost than the clerical staff of the past. The cost of this burden has been growing dramatically. An economics outfit in Australia calculates that for the country's economy, the entire gain from reducing clerical staff through automation over the past decade is being more than offset by the addition of new higher-cost checkers and trackers.

Blockchain Possibilities

Blockchain developments are now prominent in the media because there are so many places in society where distributed ledger technology can create new business models, change cost structures, and promote different behavior. The oil and gas industry offers a particularly compelling opportunity to leverage blockchain because counter parties almost always have low levels of trust and the stakes are very high, given the value of the commodity and its poor traceability. Pressures to reduce cost and improve productivity are intense.

Consider land transactions. Oil and gas companies need to acquire rights to access land to prospect for, explore, appraise, and then produce. The counter parties are often individual landowners who have limited exposure to oil and gas and may be outgunned in land dealings. There's even a phrase—"shady land deal"—to describe this scenario. Blockchain could be used to verify and eliminate fraudulent land dealings.

IN MANY COUNTRIES around the world, land records are often missing or are based on an oral tradition. In cases where the records are centralized, it is not unusual for an incoming government to station new leadership in the land registry and adjust selected land records to shift the ownership from rightful original owners to insiders.

Product sales and marketing will benefit from blockchain. Oil and gas are sold in large volumes and as such entail significant value, not unlike the size and scale of transactions between banks. The frequency of transactions is also high. A 300,000-barrel-per-day oil refinery will need a delivery every week from a very large 2-million-barrel crude carrier to maintain adequate volumes, and those cargos can cost as much as $100 million (2 million barrels at $50 each). Oil companies also need to be aware of who they are sourcing from—some exporting nations are from time to time under sanctions to prevent trade in this commodity. Blockchain can simplify oil sales and validate origin in the same fashion that it is facilitating banking transactions.

Joint ventures (JVs) are a classic example of a contract between low-trust parties over an asset, such as a producing well. The ventures are stitched together with complex agreements that need to withstand the ups and downs in the industry and are the basis for sharing the venture's costs and revenues. Most have clauses in them giving the parties the right to audit each other to ensure compliance with the contract. All parties to the JV maintain their own differing sets of records, which give rise to disputes about royalties, cost allocation, and revenue distribution. Recording costs and revenues on blockchain will ease this burden.

Blockchain technology will help oil and gas companies access valuable analytic tools in clever new ways. For example, AI software and computers for conducting various kinds of analysis are already available in the cloud, rentable by the task. The rental contracts, consisting of charges for accessing the computer hardware and the AI software, exist as smart contracts on blockchain. No humans are even involved.

Geologists who work in routine, conventional analysis may soon be able to access proprietary software for specific analysis needs without having to purchase licenses.

Blockchain technology could also be applied to facilitate data purchase and sale (seismic data is a digital asset, like a digital image). The high price of subsurface data reflects both the cost of collecting that data and the risk that the data may be exploited beyond its original sale. In the future, oil and gas companies may protect their subsurface data using blockchain and monetize it much more vigorously than possible today.

Case Study #5: Trade Settlement

Petroleum trading in Europe is about to be overhauled using distributed ledger technology. In today's world, traders negotiate buy and sell deals by phone and record the details in their separate systems (or ledgers). Paper contracts are prepared and, in many cases, mailed or faxed between parties, or perhaps converted to PDF and emailed. Barges are chartered for carriage, again by phone, and again contracted with paper documents. Inspectors are hired to verify the cargo. Other parties to the transaction may include tank farm operations, port authorities, and canal operators, all of whom maintain their own separate records and ledgers, all of which need to line up to settle the transaction end to end. Many manual systems and counterparties create plenty of room for errors and disputes.

Blockchain will introduce many efficiencies in this process, confirmed in a recent field trial. When the traders agreed to the terms of the deal, the key data was written to a blockchain, where it was recorded as a single transaction accessible and shared by the two parties, which cannot be later disputed. Any differences between the buyer or seller systems and the blockchain was their own fault—the single truth was on the blockchain. All of the other specific events (the contracting of the inspector, the barge charter, the canal passage) were likewise written to the blockchain as single agreed records of truth and connected to previous blockchain entries involving the same cargo. Unlike in blockchain, in today's trading systems, contract data can be overwritten on either side, creating the potential for a dispute.

Much of the paperwork, which encoded the key data (like port of loading, port of discharge, dates, volumes, and prices) into legal contracts, was demonstrated to be redundant with the agreed-upon and shared blockchain data. As the blockchain logged the various transactions, smart contracts triggered the next steps in the process, including launching inquiries for services (like barge and inspector services), issuing key documentation, and releasing funds.

Early participants in these trials have concluded that core industry documents, such as the lowly invoice, were now fully superfluous and no longer served any meaningful purpose. Smart contracts on blockchain could handle payment.

This trial involved just one commodity and one trading market, but the two oil companies involved estimated that a blockchain solution for finished products trading would eliminate up to 50 percent of their back-office costs (which cover managing all the paperwork, reconciling the accounts, and handling disputes). This is not a trivial sum of money in companies that often claim two back-office team members for each front-office trader, not to mention the redundancy of the admin teams in each of the parties involved (ports, inspectors, and so forth).

The companies concluded that disputes related to contracts would be almost eliminated. Business processes would become nimbler and more responsive to change. Processes would be largely automated. Cash flow would be accelerated since the smart contracts would release payment as soon as key events occurred, not within several days or weeks when the paperwork caught up.

The companies involved in this field trial have gained critical insight into how blockchain behaves, where it works, and the hassles of adoption. This small trial involved eight companies (buyer, seller, barge outfits, tank hubs, and inspectors), while a full solution would include hundreds of players. Considerable energy will be invested to pull parties together to work in this new common way. Benefits must be on offer for everyone or the solution will be hard to get off the ground. In oil and gas, barriers include antitrust laws and the scale and size mismatch between the players in a blockchain. Many consortiums will fail, but blockchain without business collaborators is like having a great Facebook page but no friends.

Initial efforts in forming a blockchain consortium will involve leagues of lawyers. Expect a lot of prework in the form of agreements on privacy and security before any tangible results emerge in real value areas. As consortiums grow, technical performance will become a consideration: selecting the right underlying technology for the process will be important.

Blockchain adoption, as with many technologies, introduces change-resistance problems. For example, in the case above, the team built the technical blockchain solution in just a few weeks, but in the end, the back-office teams could not quite bring themselves to abandon the invoice process (invoices trigger payment). CFOs and CEOs are going to have to sponsor such initiatives to get the businesses on board.

Case Study #6: Consumer Vehicles

A German automaker has wrapped up some trials integrating blockchain technology into its sports car line. This digital development is important for the oil and gas industry because it signals how consumers of fuel are being trained by the car industry, and how cars may interact with fuel infrastructure like gas stations. It also provides some inspiration for other industrial applications.

Porsche, the iconic German carmaker owned by Volkswagen (which is the largest carmaker in the world, second at times to Toyota), designs and builds cars at the leading edge of German automotive engineering. If there's a brand to trial some edgy technology, it would have to be Porsche. Its customers are refined and demanding; it sells only a couple hundred thousand cars a year, which is not a lot by global standards; and it has a heritage as an innovative racecar-maker that develops and refines car technology. Its innovations eventually roll out across the rest of the Volkswagen marques, including Audi, Bentley, Bugatti, Lamborghini, SEAT, ŠKODA, and, of course, Volkswagen.

Porsche has been running a pilot using blockchain technology with the Panamera sedan (a four-door saloon-style car); it was the first car company to turn a car into a node on an operating blockchain. Put a car on a blockchain, and a number of new features for cars suddenly come into view:

- **Accessing:** drivers can use an app to open the car or give permission via an app to another party to access the car—to retrieve something from the trunk or glove box—without giving them a key to start the car. This could be useful for delivery of fuel to the car or delivery of a package to the trunk.
- **Charging:** for the next generation of hybrids and electric plug-in cars, blockchain can track power-charging across charge points and electricity suppliers.
- **Servicing:** blockchain can keep track of vehicle servicing, like fluid change, alignment services, and insurance events. Add in smart contracts, and these services can be automatically paid for after they are executed.
- **Title transfer:** with vehicle registration recorded on a blockchain, the data about the vehicle moves to the next owner more swiftly and with fewer errors.
- **Data sharing and learning:** blockchain can hold data about driving routes and preferences, which in aggregate helps shed light on customer behavior.
- **Costs:** blockchain apps enable payments for parking, via a smart contract. Similar structures could be used to pay for fuel, make car loan or lease payments, pay for insurance, or even settle fines.
- **Rentals and sharing:** smart contracts on blockchain enable vehicle-sharing for journeys or space. A compound of residential units now shares a common fleet of rental cars, using blockchain to record usage and settle accounts.
- **Insuring:** paying drivers via blockchain with tokens or loyalty points for good driving behavior (no idling, hard braking, or fast starts) enables gamification.

Toyota is also trialing blockchain, with an eye on enabling a future autonomous vehicle world. I can imagine robocars easily enough, but one with the smarts to navigate our human-centric automotive world will need something like blockchain to enable all the interactions that I've set out.

Blockchain's Future in Oil and Gas

A car on blockchain is a rich example of turning an industrial asset into a full participant actor on the Internet and giving it a kind of agency for its many interactions with society. There are many other kinds of assets in oil and gas that could also benefit from being nodes on blockchain, becoming participants in the energy economy and gaining agency. These assets will not have anything like the rich set of interactions that cars have, but the volume of interactions is still meaningful, and the high levels of trust placed on these assets creates a pathway for agency.

A linear extension of the car-on-chain would be pumps and motors on blockchain. Aggregated condition data is probably less valuable than, say, customer-journey data, but for some pumps and motors, choosing when to run matters. My condo building has a large fan that cycles when the heat buildup in the atrium reaches a certain level. What if that motor cycled based on the source and price of electricity, delaying when prices were spiking? Using a smart contract structure on blockchain could make the motor much cheaper to run.

The oil and gas industry has a lot of meters—at wellheads, on pipelines, and in tanks. Any time product moves through the assets, it gets measured, and the measurements are used for title transfer, insurance, contract fulfillment, quality control, pricing, and so on. The measurement devices are tightly controlled, and the meters are regularly calibrated because the data is so valuable. Meters could become nodes on blockchain and record measurement data to blockchain where it could be used in smart contracts.

The industry is adopting drones for subsea inspections, piling monitoring, aerial surveillance, pigging, tankage inspections, and so forth. These robots are no different from the robocars coming to the roads—most are going to be shared assets operated by field services companies, with the opportunity to apply smart contracts to their usage. For example, an aerial drone that measures progress on a construction site could trigger a smart contract payment to contractors based on visual data about progress. That drone might be a new kind of field service.

The industry purchases a huge amount of new equipment for its projects, some of which is never actually used, resulting in a buildup of surplus asset inventory. In the same way that cars will use blockchain for title transfer, oil and gas could put its purchased assets on blockchain for their eventual sale.

COMBINING DIGITAL TECHNOLOGIES

While these individual digital technologies have impact on their own, some truly wondrous results accrue when they are used in combination with each other. For example, Airbnb brings the power of smartphone users and cloud computing to create a new accommodation-as-a-service solution, while, behind the scenes, artificial intelligence is processing historical visitation data to help find the best rental for the customer, to set prices, and to recommend additional services based on customer history and available options. Netflix uses cloud computing to stream content to its customers' devices, synchronizes the customer profile to make it globally portable, and leverages artificial intelligence to organize individual menus of content based on prior viewing history.

There are already useful examples in oil and gas that demonstrate the incredible impact combinations of digital technologies can have on the business.

Case Study #7: Royalty Processing

A Calgary–based oil and gas company working with other oil and gas operators, a bank, and a royalty specialist has created an oil and gas royalty smart contract on distributed ledger technology and executed a payment based on the smart contract. Some view this innovation as much ado about nothing—smart contracts are novel but not really new, revenues did not improve, and certainly innovation in the back-office in oil and gas attracts little more than yawns from the engineering world. But look a little deeper.

THE CHALLENGE

The oil and gas industry generally operates under some kind of hydrocarbon royalty- and revenue-sharing regime. For those not familiar with this aspect of the industry, royalties are a kind of economic rent paid to the parties involved in hydrocarbon development and extraction. Revenues are shared with other participants as well. Take, for example, the case of a petroleum well located in a farmer's field. The rural landowner will expect to be compensated for the inconvenience of having the oil company continually access their land. The well pad may no longer be used for grazing or cropping, and trucks and rigs need rights of way for tracks and lay-down areas. Accordingly, the farmer is paid a share of the profits that come from the well's output. In Queensland, Australia, these pads are the single most valuable acreage on the farm based on cash proceeds per square yard.

Wells may have more than one owner, as in the case of joint ventures involving two or more oil and gas companies, and over their lifetime, wells may be sold wholly or in part to other parties. To add to the challenge, wells can be in operation for decades, handed down from one management team to the next. The land on which the wells are located may also be transferred as inherited property and sometimes split across multiple family members. The land can be sold, too, bringing new landowners into the royalty equation.

Governments are also deeply involved in royalties. In Commonwealth nations such as Canada and Australia, the hydrocarbon resource is actually owned by the public, and the regional government in turn sets out the government royalty calculations. In the U.S., the landowner owns the rights to whatever is extracted under their land, so royalty calculations are a matter of negotiation between the landowner and the oil and gas company. Internationally, governments set the royalty rates for resources extracted in their territories.

The algorithms to calculate the royalties owing to the various parties are complicated. Sometimes governments want to stimulate development of specific resources in specific areas for specific time periods, and so the royalty rates may be reduced. Sometimes certain costs are deducted from the royalty calculations in an accelerated fashion to stimulate economic activity. When governments change, they often can't resist tinkering with the royalty rates to extract more revenue from the resource. Finally, it is a challenge to anticipate all the different costs that could be incurred against the well and the rules for sharing those costs.

In any case, the parties to the well get together and agree how the revenues and costs are going to be divvied up among them and write up the sharing agreement in the form of a contract. The contract then becomes the basis for determining who is owed what. In international settings, the contract might be called a production sharing agreement. Oil and gas companies must then employ an army of back-office production and royalty accountants, lawyers, and systems professionals to figure out the royalty calculations each month.

As you can imagine, processes and systems for determining royalties are prone to error. There are hundreds of thousands of wells, paper contracts, and corresponding Excel spreadsheets that need to be maintained. The formulas in the spreadsheets are frequently wrong, out of step with the contract, or no longer current with the prevailing rates and rules. The parties all try to maintain their own separate records, the effort is largely manual, the calculations are monthly, and the calculation formulas change frequently. Larger players might have their own dedicated systems, but that's unaffordable for many smaller companies and certainly for individual landowners.

Since royalty calculations can be easily manipulated, royalty holders suspect that well operators may be exaggerating the costs and minimizing the value of the production, which lessens the royalty payment owed. Operators may delay the royalty payments to manage cash flow. Suffice to say, the industry is rife with disputes, which incur needless legal fees, the occasional court challenge, payment delays, and ever more complex agreements.

Unfortunately, the system is highly resistant to change. In early 2017, I met with the senior vice president of a leading oil and gas outfit to discuss the impact of commodity pricing during the downturn, by that time well into its thirty-sixth month. He was quite proud to point out that while his company as a whole was well into its third restructuring, there had been no impact at all on his finance and accounting teams. These manual processes were still intact.

THE SOLUTION

Distilled down, this business problem features a huge volume of English-language legal contracts that require specialists to interpret them correctly and low levels of trust between counterparties.

The first challenge was how to replace the production accounting, royalty, and contract expertise whose role was to read these contracts and interpret them. Natural language processing is a specific artificial intelligence skill that has now reached a level of capability that makes it suitable for this problem. The various contracts, which existed as text documents in paper or PDF form, were fed into an AI engine that interpreted the contract terms and conditions using natural language processing and extracted those terms that are the basis of the royalty calculation.

Of course, the AI engine must be "taught" how to read and interpret the contracts, which requires the initial help of specialist royalty experts. There is a lot of upfront work involved, but in time the AI engine improves its performance. In fact, in tests, IBM's AI engine Watson outperforms lawyers in both speed and accuracy in reading contracts and identifying contract language with errors and issues.

Once the key terms from the royalty contract were extracted by AI, the next step was to translate the terms into an algorithm to calculate the royalty. This is broadly a mechanical exercise of building a spreadsheet with the correct rows, columns, and cells with labels and formulas. This task was handed to a digital toolset called robotic process automation (RPA), which is software that basically mimics human keystrokes in carrying out repetitive work. The RPA tool then built the algorithms and calculations that would calculate the royalty.

The third innovation was to turn the royalty payment calculation into a smart contract on distributed ledger technology. Once all the data is available, the royalty is calculated and the smart contract issues payment to the royalty holder automatically. In this way, the participants to the royalty contract can see exactly how the royalty was calculated, based on the original royalty contract, and can eliminate their duplicative efforts to calculate the royalties.

Now that this innovation is in place and paying off, it's easy in hindsight to see why oil and gas companies, landowners, governments, and regulators might want to modernize how royalties are handled.

First, this digital innovation is helping to eliminate many of the disputes that plague the royalty area. Disputes have been long seen as a cost of doing business, but with low oil and gas prices, and in times of economic contraction, these costs loom larger on a relative basis.

Second, this innovation is helping to accelerate cash flow in the industry. Payments may be delayed while a dispute is outstanding, so elimination of disputes accelerates cash flows for the royalty holder. Smart contracts execute automatically when conditions are met, and the practice of "slow-walking" the royalty payment to the landowner is also largely eliminated.

Third, there are two kinds of cost savings from this innovation. Fewer disputes means lower administration costs for both the operator of the well and the royalty holder. Both parties gain. The second cost savings comes from reducing the monthly effort to calculate the royalty, which is automated away. Reduction of both costs means the royalty value should go up, which should please governments and royalty holders.

Finally, the industry is once again made nimbler. When royalty rates change, as they do on a frequent basis, the AI engine, the robotized-spreadsheet machines, and the smart contracts on blockchain will be much easier and faster to adapt than armies of production accountants and mountains of spreadsheets. One company estimated that its royalty calculations following a rule change were typically incorrect for months, creating a backlog of catchup activity. There will be a longer-term reduction in the uncertainty and volatility in royalty values. Investors should like that.

This example should trigger fresh interest in several other similar areas of oil and gas business that also rely on expensive human capital to read, interpret, and act on contracts. Royalty calculations are just the first of what could be a wave of artificial intelligence, robots, and blockchain applied to oil and gas. Other areas include services agreements, land and lease agreements, supply chain contracts, procurement, and purchasing contracts.

THE OUTCOMES

As was outlined earlier, the oil and gas industry has already started to upgrade its ERP platforms to the next and more digital versions.

However, this example demonstrates how some business processes typically implemented in an ERP system could be quite different in a world with greater digital innovation. For instance, smart contracts on blockchain may not require invoicing, receivables, and payables processes, three big financial activities found in every oil and gas finance department. ERP deployments could capture significant new value with digital enhancements, and ERP implementation teams really should step back to understand how digital innovations could transform business processes before applying ERP to them. Executive sponsors need to be challenging ERP teams about how blockchain technology coupled with machine learning could transform finance, HR, supply chain, capital projects, and maintenance.

This royalty example also points out again how digital technologies are impacting what are considered high-end, privileged, and well-paying jobs. In many oil and gas companies, the legal, accounting, royalty, and land jobs are held by women, a workforce segment that is already underrepresented in the sector workforce. The future of work in oil and gas companies is going to look very different. Human-capital leaders in oil and gas need to get ahead of these technological developments to help cushion the blow to head count. CFOs, in particular, need to prepare their organizations to embrace these new technologies. The future of finance should feature a lot more smart machines.

Case Study #8: Gas Factory

A second example of how a combination of digital technologies can transform oil and gas comes from the upstream world of well-drilling and asset delivery. For many years, since at least 2006, upstream oil and gas companies working with unconventional resources have speculated about becoming more like manufacturers and less like traditional oil and gas explorers and producers. Manufacturers have lower cost structures as they can take better advantage of purchasing efficiencies, scale economies, and overall optimization. A stark example comes from the coal seam gas industry in Queensland, Australia, where the gas companies are drilling 10,000 gas wells to feed the state's huge new LNG export industry. Viewed through standard industry practice, each well would be individually designed, specified, drilled, completed, and connected to the gas-gathering network at high individual well cost. However, an executive with one of the coal seam gas companies pointed out that his company should not shape the business around the task of drilling 10,000 wells—it would drill just one well and then it would repeat that effort 9,999 times, like a manufacturer. His company and others had been searching for the ways and means to achieve this goal.

THE CHALLENGE

Despite initial efforts to change the equation, costs for well delivery had been frustratingly sticky, tending to rise in lockstep with commodity prices and fall only under the most difficult of circumstances, such as when there is a major market correction like the collapse of oil prices in 2014.

This is not how manufacturing costs behave. Manufacturers watch their costs very carefully and constantly invest in ways that expand their margins (the delta between their input costs and the price they can command for their products). High variability in controllable costs is a sign of poor manufacturing management. If oil and gas companies want to get their costs to behave the way they do for manufacturers, they need to do more than harangue their suppliers for cost reductions. They will need to adopt at least one other key technique that has been the hallmark of the manufacturing industry since the days of Henry Ford—the assembly line.

Back in the early days of the auto industry, cars were built the same way that oil and gas wells are built today: by craftsmen. The car, like the well, didn't move in the shop. The craftsman came to the car with their tools and parts and slowly assembled the vehicle over time.

It was hard for the craftsman to cut their costs. Most of their value was in their experienced labor. Parts were built by hand. Expanding production was a problem because experience takes time to accrue and experienced labor was scarce. Labor rates escalated dramatically when demand grew. Since their tools had to be portable and human scale, they were small and hard to finance. Portable assets tend to walk away and are hard to pledge as bank security. Substituting individual machines for individual people was impossible.

Production volumes were small and making the assembly more efficient over time was slow and dependent on the skill of the craftsman. Even today, high-end sports cars with limited production volumes are made mostly by craftsman in assembly shops. Other products—like bespoke shirts, buildings, and art—continue to be built in this craftsman model.

Oil and gas wells are the same today. Fixed in place, the well site is in effect the assembly shop. We might call it a well pad, and it might have a water-handling facility, a lay-down area, parking for vehicles, and tanks for production. The craftsmen are the service companies who go to the well site with their teams and equipment (rigs, pumpers, and frac spreads) and carry out a service on the well, either to build it in the first place or to repair or service it later. A simple gas well will be serviced by dozens of different companies and involve dozens of different work steps to bring it online.

As with all craftsmen, service companies struggle to do much with their costs. Labor rates are sticky and don't easily retract when the market tightens. When the market expands, the suppliers have to pay more for labor and their costs go up. They can also raise prices quickly when demand exceeds supply and when oil and gas companies relax their focus on costs. Their assets are, for the most part, financed by bank lending, which demand regular payments; with demand uncertainty, suppliers charge what they can when times are good to make up for when times are not. They can innovate a little with their rigs and kit, but those efficiencies are at best incremental to specific tasks and not transformative to the overall asset life cycle.

Henry Ford changed manufacturing forever with the invention of the assembly line. The assembly line helped drive down costs by replacing craftsmen with less skilled labor and simplified tasks. But the big investment by Ford was in creating the mechanical conveyor system that moved the vehicle past assembly workers who were arrayed in workstations. The forward work queue for the assembly worker was defined by what the conveyor system delivered to them. Delays caused by missing parts and missing workers were eliminated. The line could be sped up or slowed down, or run round the clock, to match production to demand.

The production volumes achievable in today's auto plants (up to 500,000 units per year per assembly plant) are simply not possible in a plant whose basic operating model is craftsman-based assembly.

Unfortunately for oil and gas, the well, the flow lines, and the batteries do not move and will not ever move. They're stuck in place because that's where the hydrocarbons are located. So, the industry cannot simply copy Henry Ford's innovation by locking the service companies in place and bringing the well to them on a conveyor belt. This basic physical limitation means no option but to bring the craftsmen to the well, using the road and highway system.

THE SOLUTION

The coal seam gas operators recognized that the industry needed to convert the work to be done on wells and facilities into a virtual assembly line rather than a physical assembly line. One company hired several process engineers from the Japanese auto industry to help design the virtual assembly line to enable the development of a Gas Factory, a business that would manufacture gas wells like a car company makes cars.

Three digital tools working together provide the visibility needed for the work to mimic an assembly line.

- Mobile devices (smartphones and tablets) provided instant access to information on work to be done on wells and infrastructure.
- Cloud computing made a collaboration system accessible to the dozens of participating companies, the landowners, the regulators, employees, and it housed the data about work to be done, the wells, and the suppliers.
- Analytics were employed: algorithms adjusted work queues for the service companies as conditions changed.

Service companies could now see a work queue stretching out in front of them in the same way that an assembly-line worker can see the vehicles coming at them in a factory, instead of the one-at-a-time purchase orders of old. They could access the work queue while on the road on a smart device, rather than only back at an office. Well owners could see the sequence of services to be carried out on their assets and could make more thoughtful decisions about accelerating some work and slowing down other work, based on real data. Working together, service companies and asset owners could optimize the performance of the logical assembly line.

The benefits of this kind of digital change have been profound. Well delivery improved from one well per week to seven wells per week. Suppliers of parts got in on the act, too, eventually packaging thirty-two wellheads into a single container, versus an initial design of one wellhead spread over two containers (a sixty-four-fold improvement).

THE OUTCOMES

While there are several new digital solutions that help improve work management in the field, most do not offer the comprehensive and fundamentally new business model of the virtual assembly line. Other changes to the operating model are necessary.

- The relationship between an oil and gas company and a service company needs to change from one of adversaries to business partners. Toyota does not treat its just-in-time suppliers the way oil and gas companies treat their suppliers. Auto parts companies are much more involved in open and joint business volume planning.
- The design of well infrastructure would benefit greatly from some further standardization, which is more achievable in the unconventional shale and coal seam gas world than in the conventional world. Standardization helps capture service efficiency and cost reductions.
- Oil and gas companies need to think of their asset data as powerful enablers. It is not in the nature of an oil and gas company to share its planned work program openly with its suppliers.
- Both oil and gas companies and service companies need to bring on different skills, including more manufacturing process expertise from the likes of Toyota, Boeing, and IKEA.
- Performance measures need to adapt to reflect a manufacturing mindset. I would anticipate greater emphasis on the measurement

of task time, travel time, on-time performance, and quality performance on task.

- Technology choices would need to be more enterprise in nature and less field-specific, asset-centric, and driven by narrow business unit agendas. Manufacturers might let the paint shop toy with new spray nozzles, but they still need to be part of the assembly line.
- Oil and gas companies would need to reorganize to create more of a controlling function that looks for ways to optimize assembly activities, based on demand for services, constraints in delivery, such as inclement weather, and travel distances.

These are harder changes for oil and gas companies to deploy because they get at the very basis for how oil and gas has long organized itself. Adopting digital tools in this aspect of the industry allows for a fundamentally new way of well delivery at scale to emerge, which is increasingly important as the industry shifts to more and more unconventional resource plays.

KEY MESSAGES

It's becoming clear that outsized value comes from combining digital technologies in new and creative ways to unlock innovative new business models. Executives of oil and gas companies, suppliers to the industry, and regulators need to anticipate that digital solutions will become a disruptive force sooner rather than later. However, moving forward on a digital transformation isn't straightforward—the way other industries innovate using digital ("Run around and break things," to quote a wag from Silicon Valley) will be rightfully rejected in oil and gas.

Here are a few takeaways from this summary of digital technologies:

1. Cloud computing will store the new flood of data, enable new disruptive business models, and provide the foundation for most other digital innovations.

2. ERP systems will continue to provide the commercial underpinnings for the industry, while becoming more digital in their design and operations.

3. Artificial intelligence will read and interpret all the data, supporting key human decision-making functions.

4. Sensor technologies and the Internet of Things will unlock remote asset monitoring and maintenance and process efficiencies, while generating vast quantities of highly valuable data to store and analyze.

5. Autonomous technologies will apply the data and analysis to execute work, replacing humans in the office and the supply chain, as well as in costly, dangerous, and repetitive work. In the process, entirely new jobs will be created.

6. 3-D printing will transform most other industries and their supply chains and lead to reduction in oil demand.

7. Digital reality will become the design and execution arena for humans to engage with the physical world in new, rich ways.

8. Gamification will enable workers to study, experiment with, and play with their actual businesses and industries in more profound ways.

9. Blockchain will transform business processes involving assets, trust, ownership, money, identity, and contracts.

LONG FUSE, BIG BANG: Digital's Impact on the Value Chain

GROWING UP, I learned to appreciate the comic antics of Bugs Bunny, Elmer Fudd, and Yosemite Sam. One frequent gag involved a powder keg and an absurdly long trail of gunpowder, usually lit, which the hapless Elmer or Sam tried to outrun. It didn't seem to occur to either of them to simply throw the keg away; instead, they always ended up on the wrong side of an epic, non-terminal explosion.

The size and timing of digital implementation's impact is analogous to a keg of gunpowder and the length of the fuse. The bigger the keg, the bigger the bang, and the shorter the fuse, the more quickly the bang will happen.

Among those who count themselves as early digital adopters, there's a certain breathlessness and hype about digital's short fuse; they assert that it's already here in the form of smartphones, drones, video streaming, and so on. The retail sector is clearly feeling the effects of Amazon's online shopping model. The banking sector, particularly retail banking, has been moving very quickly to create smartphone apps to usher in a new banking experience. In this way, retail and banking are feeling the effects of digital sooner rather than later, and the impacts are huge, including new and highly disruptive business models. Short fuse, big bang.

But the fact is that of the world's currently installed oil and gas production and distribution capacity (producing wells, crude oil pipes, plants, and refineries), the vast majority (80 to 90 percent) was built long before digital was a concern. They are ill-equipped to participate in a digital world without a hardcore retrofit, which is tricky to pull off since these assets run 24/7, are highly regulated, and are frankly dangerous to people and the environment when they misbehave. Retrofitting digital onto an existing long-life operating asset would be like installing a pacemaker on a marathon runner in the middle of a race. It should be no surprise that oil and gas will adopt digital much more slowly than other industries.

For oil and gas, the digital fuse length is going to vary depending on the dimension of the value chain being impacted. The size of the bang will also vary, with digital causing tremendous upheaval in some areas and relatively modest effects in others.

But the bang will be big in the aggregate—there is now simply too much money flowing into the transportation sector alone for it to be anything but big. At the moment, it seems unlikely that an entirely new model of the oil and gas industry will unfold through digital innovation, as retail has been transformed by the online model. But there are already clear signs of new business models emerging in fuel retailing that are potentially disruptive to the incumbents. The sharing economy, specifically for field services, could be viable. Tolling models for oil and gas facilities beyond the LNG plants could take off. A business model that separates the ownership of the intellectual property of chemical recipes from the ownership of processing plants could be brought from the pharmaceutical sector to oil and gas. Perhaps in a few years' time, and with the benefit of hindsight, analysts will be able to point at key tipping points in the drive to digitalize.

A client of mine once counselled that management teams and boards typically want to know, within reason, answers to a number of questions before they're prepared to commit resources to a course of action. Digital investments will receive the same scrutiny as other investment proposals. Here are the kinds of questions I typically hear around the board table:

- When will the impacts of digital be felt?
- What will the impacts actually be—good, bad, or both?
- How consequential will the impact be?

- What are the consequences of doing nothing?
- What are our peers doing?
- What technologies are making a difference?
- Who is the leader in adopting digital innovation, and what are they doing?

These questions are not confined to upstream companies or field services firms; they're being raised across the entire sector, from one end of the value chain to another. The table below projects the timing and impact of digital on the sector.

Timing and Impact of Digital Innovations on Oil and Gas

	FUSE	BANG 1 - low, 5 - high
Exploration	1–5 years	5/5
Production	5–10 years	2/5
Field Services	1–3 years	4/5
Processing and Refining	5–10 years	1/5
Fuel Retailing	10–20 years	3/5
Capital Projects	10–20 years	4/5
Turnarounds and Overhauls	5–10 years	2/5
Support Functions	5–10 years	4/5

The previous chapter explored in some depth several specific digital technologies. This chapter reviews the oil and gas value chain and how digital change will impact the future of the industry.

EXPLORATION: A SPECTACULAR BANG!

Fuse length: one to five years
Impact: 5/5
Key technologies: artificial intelligence, machine learning, cloud computing

By far the most data-intense function in the oil and gas value chain is exploration. The exploration function of oil and gas has always been able to attract the funding and technology investments required to collect and interpret vast volumes of data, which help to characterize geologic formations and identify the possible presence of oil and gas. After all, the upside can be highly lucrative if the exploration is successful. The upstream departments of most oil and gas companies have plenty of rich datasets at hand and plenty of analytic support, but most will readily admit that they are not particularly digital.

Exploration's Digital Paradox

Despite being extraordinarily data- and compute-rich, exploration teams tell me that they trail in the adoption of digital innovation. Using my definition of digital as the combination of data, analytics, and communications, I wonder why they would classify themselves as digital laggards.

Let's begin with data. The data volumes in oil and gas (such as seismic data) are enormous, but, historically, data transmission speeds have been too slow to move the data along communications networks for analysis. As a result, and still to this day, most exploration departments operate their own private data centers so that the datasets do not have to move. This has saved time for geologists and petroleum engineers in the past and has given the larger players, who are able to afford big data centers, an advantage.

Oil and gas companies treat all of their geologic data as highly proprietary and deeply competitive. This makes sense when the value of resources depends on the data. Indeed, there is a lively market in the trade of subsurface data. In the main market, however, the tendency has been to keep this kind of data in a company-owned data center under lock and key. The silos of data inside some exploration departments could lead you to conclude that internal competition is in fact greater than that with others in the industry.

Exploration departments are demanding users of analytic horsepower and have been very compute-intense. Cloud computing should be very appealing to them, but thus far it has been embraced only tentatively, and in general not within the upstream. I believe this reflects an ongoing bias and long-held orthodoxy that third-party computer installations cannot be competitive, even at scale with the captive compute facilities in oil and gas companies.

In a meeting with me, the CIO of one of Canada's largest oil and gas companies dismissed cloud computing as a rent-extracting monopoly to be avoided. The industry has been wary of creating monopolistic dependencies in its business model. It likes to use multiple suppliers for any one service, prefers multiple pathways to market, and invests across multiple jurisdictions. Putting data into a third-party data center (via outsourcing) or into a cloud-computing environment allows the cloud provider to jack up prices to its now-captive oil customers. My CIO contact is convinced that well-run oil and gas companies can match cloud-computing companies for capacity and cost.

What about the use of new digital tools like artificial intelligence? Most geologists simply do not accept that the interpretation of geologic data, which involves marrying subsurface imagery and geology training, can be done by machines. Even suggesting that AI has a role to play attracts derision from the industry and accusations that the technologists don't get the industry.

In January 2018, I was presenting at a conference in Lake Louise, Alberta, on the impact of digital on oil and gas. I had just stepped down from the podium after asserting that AI would unlock more reserves, possibly without too much human involvement, when I was met by an irate geologist intent on correcting my perspective. His view, presented with the authority of religious dogma, was that digital cannot and will not ever match the geologist for data interpretation and that I was foolish to even suggest it.

This does not square with what we are observing in the world around us. Faster growth and better business valuations are associated with companies in the digital world which treat data as an open and shareable asset and which leverage compute on demand. Artificial intelligence is giving us self-driving vehicles, energy optimization, and data analysis. AI is interpreting medical imagery such as ultrasound images (which uses the same technology as seismic data capture) at a performance level that matches and often exceeds human experts.

Threats to the Status Quo

As time marches on, the status quo position of building and operating a proprietary data center inside an oil and gas company begins to look dated and even risky. It is doubtful that any single large industrial company can effectively match its compute facility cost to those of the cloud-computing companies. Size, scale, power needs, and automation levels suggest the digital players have the upper hand. The cloud providers are likely to offer better cyber security, too; the protection against an attack on one of the handful of big cloud-computing firms can be more quickly shared to its peers than to the hundreds of thousands of independent small data centers.

The sharpest skills in compute technology will blossom in the embrace of the cloud-compute operators. Oil and gas has always been able to attract its share of computer science majors, but, to many graduates, the new digital economy now looks more appealing.

As the world's software products migrate to the cloud-computing model, companies with a legacy investment in proprietary in-house software will increasingly find themselves with a growing stockpile of dated, unmaintained software. The most innovative, creative solutions and technologies to modern problems will not be available to those organizations wedded to their own infrastructure. Business risks will start to rise.

Alternative Approaches

A few industry players have started to explore if this model of proprietary data, on-premises data farms, and roll-your-own computer centers is still the only way to go.

In mid-2017, I was invited to attend a workshop hosted by the IEA to discuss the impact digital would have on the energy industry. The workshop included representatives from a number of large oil and gas companies, one of whom presented their experience in trialing digital thinking in exploration, and the results were startling and dramatic.

This large global oil company set out to test the current state of big data, computing, and analytics to see if the digital capabilities available on the market could meet its demanding performance goals. The company loaded a sizeable volume of seismic and other subsurface data to a commercially available cloud-computing environment (from one of the market leaders) and invited a cadre of data scientists and algorithm specialists to analyze the data using whatever math and tools they had at their disposal. The data had been anonymized so that its competitive value was neutral, and it was presented in such a way that non-industry professionals could work to solve some of the problems the company wanted to study.

The company concluded that the world of commercial cloud-computing services is now sufficiently robust that it could take on its demanding data volumes and analytic challenges. This is a profound change from the status quo. In the not-too-distant future, oil companies will start to question the need to maintain their proprietary in-house data centers. These facilities are often located in some of the most expensive office real estate in the world's costliest cities. By parking the enormous quantities of data in the same cloud facilities that house the analytics, exploration departments can also sidestep any communications bottlenecks.

The data scientists, who were drawn from many industries, were able to match the company's own geologists and petroleum engineers at interpreting the data and identifying the most prospective locations. Interpretation of seismic data is supposed to be a core competence solely of oil and gas professionals. They get paid outsized sums for their knowledge and ability to work with this data. This experiment proved that an attractive and viable alternative to creating a house level of expertise is now globally available.

Growing the Reserves

Aside from rendering in-house data centers obsolete, and creating a new crowdsourced data interpretation model, what other impacts is digital expected to have on the exploration industry? What is the likely impact on reserves?

The biggest impact by far is in the forecast for reserves and its growth rate. The IEA estimates that digital, applied to the unconventional oil and gas resources (that is, shale and other low permeability and low-porosity geologic layers), will expand global oil and gas reserves by 5 percent. It doesn't sound like a lot, but that translates to 70 billion tons or 500 billion barrels of oil equivalent. With demand at about 100 million barrels per day, this volume equates to thirteen years' supply.

While many countries in the world have shale resources (U.S., Canada, Argentina, China, and the UK), so far only the U.S. and Canada have figured out how to extract hydrocarbons in meaningful volumes from their shale deposits. Much of this growth in reserves will accrue to these two countries. This helps explain why the IEA also forecasts that the U.S. may well become the world's largest oil exporter in the next few years.

In addition to the low-permeability and low-porosity reservoirs, the legacy conventional deposits, with recoverability averaging just 40 percent, will also benefit from better analytics and understanding of the reservoirs. Suffice to say, there's a very big prize awaiting the early adopters of a few digital techniques.

Exploration's Digital Future

The future of the exploration function in a more digital world could look very different from how it is managed and delivered today. Today's business model, which relies mostly on in-house capability, could give way to more involvement by third parties and crowdsourcing. The resource owner could arrange services to the asset, separating the role of owner from operator and from supplier in the same way that surface assets are structured. Data might be stored with one supplier, while analytics might be with a second company, and specialist human skills with a third. Multiple resource owners, with their data stored in accessible clouds and not behind private walls, could pool their data to create larger, more comprehensive data oceans. These vast datasets could be more readily opened up for third-party analysis and enable marginal players to access the interpretive resources of the supermajors. Resource owners ultimately will behave more like orchestra conductors and less like monolithic vertically integrated operations. Data-interpretation algorithms will feast on the data bonanza and improve dramatically.

Entirely new businesses could emerge in such an ecosystem. Once data scientists and algorithm specialists figure out how to interpret data, they are often tempted to codify what they have learned into software.

In a cloud-computing world, that software will be in the cloud and accessible to anyone with an interest in interpretation and analysis. Cloud-only data-interpretation businesses could thrive, be accessible globally to anyone with a resource that could benefit from fresh insight. This "data and algorithm heavy, asset light" business model could disrupt the upstream industry.

Digital tools have important applications beyond geologic interpretation. Imagine smart drill bits with sensors behind the cutting wheel that capture real-time data about the formations being drilled, sending that data to the cloud for comparison and interpretation against enormous seismic datasets and all prior drilling jobs, to help direct-drilling operations. Fleet learning comes to well-drilling, which allows smaller production companies to leverage the productivity insights from their peers.

PRODUCTION: SPUTTER AND FIZZLE

Fuse length: seven to ten years
Impact: 3/5
Key technologies: Internet of Things, analytics, cloud computing

Of the many areas of oil and gas where digital investment could be valuable, production should be a strong candidate. Production is the revenue end of the business, margins in the overall industry are usually at their maximum in production, and modest gains in volume or asset throughput pay off quickly. Over the years, production has been the target of ample investment in information tools to improve asset performance.

Despite this promise, production sits squarely in the category of late digital adoption. There are a number of reasons for this conclusion, such as economics, field age, energy transition timelines, infrastructure shortcomings, the rate of digital change, and risk concerns.

Oil and gas production is a self-destructing business model. For every barrel of hydrocarbon that a well produces, it has one less barrel in inventory, which creates a built-in pressure for producers to explore for more. Every well has a decline rate, and on average around the world, a given well produces 4 to 6 percent less than in its previous year. The cash flow from production pays for the search for more, giving managers a strong economic incentive to produce from wells as economically as possible, for as long as there are barrels to produce. The trick is to spend just enough on the job, and ideally as little as possible. Even if the demand for oil and gas did not grow, the industry would still need to replace 5 million barrels of production each year, which takes a lot of capital.

The age of a field factors into decisions about whether to invest in its digitization. Many production fields will produce enough volume that investments in digital infrastructure (sensors, reliable power, or telecommunications services) can make economic sense for a year or two. But decline rates and average production lifetimes force owners to recover their digital investment in these assets very quickly. Engineers may struggle to justify the investment to bring these older fields into the digital age, considering how long the field is expected to produce.

Of course, there are wells, and then there are *wells*. Some produce very low volumes in a steady drip way and are under the barest of life support. Others produce phenomenal quantities of the best stuff (low in sulfur and other contaminants) and garner a lot of management attention. There are oil wells, gas wells, oil and gas wells, oil sands mines, shale wells, heavy thermal wells . . . the list goes on and on.

Observers of the industry estimate that digital could improve well productivity by 10 percent or more, but that begs the question of where and how. What is it about oil and gas production, an industrial process that has been ongoing for more than 100 years, that could yield that kind of benefit?

Current State of Production

Production has four interrelated challenges, each of which could be better enabled by data. They are:

1. understanding the geology of the well and using that insight to strategically manage its performance;
2. managing a geographically complex network of production assets that can span millions of square miles or is many miles offshore;

3. coordinating and managing the services required across that geography; and
4. managing an enormous and complex data puzzle that is growing constantly.

First off, there are already millions of wells, and the industry adds thousands each year to replenish those that peter out. Keeping track of them all is a challenge in itself, especially when they are bought and sold and when the landowner is not the mineral rights owner.

Hydrocarbons are found all over the earth, so wells and their related infrastructure (plants and pipelines) are found in many countries, some with relatively weak economies. Production fields can then cover a huge land area. Queensland, Australia, is home to one of the world's largest gas fields, supplying billions of cubic feet of gas every day to the LNG sector, and its fields are about the size of Finland (if you added up the surface area of the Surat and Bowen basins). Logistics to service all these wells can be costly. Most wells will operate for many years before they eventually run out, which can create a lot of data about the well performance, the geology, and the services required to keep wells producing.

For digital innovation to be truly valuable, it needs access to telecommunications networks and reliable power supplies. Telecommunications infrastructure is frequently weak in remote locations, and individual wells and fields may be too marginal to cover the cost of installing high-capacity networks and mains access. Unless the expected volumes are high, producers will struggle to make the economics of digitization work.

It also turns out that there is exceptionally wide variance in the wells, their designs, and their specific equipment on-site. To date, there are no strict standards in well design and equipment configuration, even within single companies or fields. This is in part a legacy practice from conventional oil and gas, where wells were always bespoke design (tolerable when margins are high, or production rates are substantial), and in part due to uncertainty around best practices. Engineers like to design, after all. Service protocols vary by well, the spare parts are probably more diverse than optimal, the tools needed to maintain them are also varied, as are the skills required to maintain them. Managing this kind of diversity is an information problem—the better the quality of information about the well, the less of a challenge this variation poses. But producers typically have poor information about these assets and consequently struggle.

In this industry, field services are generally purchased from many different service providers from small local companies to large sophisticated outfits. Smaller service providers tend to concentrate on their area of capability, offer a restricted service range (such as electric power testing), and survive through a steady stream of service orders that send them hither and yon, from well to well.

Wells, flow lines, batteries, pumping stations, and compression assets have many different kinds of service needs, including testing, valve calibration, commissioning, inspections, cleaning and lubricating, vegetation removal, general repairs, work overs, pump servicing, water separator maintenance, and filter replacement. Frequency of service varies, and order of service needs to be considered. Some of these services can be scheduled (through predictive techniques) while others are on demand. The service company relationship is usually at arm's length, with both customer and supplier maintaining separate records about the services required and delivered. The level of integration between parties that is common in manufacturing is still foreign in oil and gas.

From time to time, there are "acute incidents" that interrupt production and create above average demand for services. These incidents include heavy rain, floods, and fires, which can take whole fields off production. Acute incidents require more coordination between rivals in a given basin and faster surge response than normal. Producers in specific fields, like the Gulf of Mexico, may find themselves all trying to draw on a limited service industry at the same time during periods of acute need, such as during a hurricane. During the fires that impacted the Fort McMurray area in 2016, producers scrambled to effectively mobilize emergency services.

Many older wells are connected to SCADA systems that provide visibility to the well. They likely each have a dozen sensors installed, and these sensors throw off a steady stream of data points, recording temperatures, pressures, volumes, and power usage. The volume of data is growing over time, along with the wells. Making sense of all the data in real time is hard enough, let alone sifting through it over time. Newer wells will have even more sensors thanks to the falling cost of sensor technology. Only cloud-computing technologies can help sift through the data quickly and cost effectively.

Production businesses can be pretty complex, with dozens of systems playing some role in sustaining the assets. These systems include resource planning, geographic information systems, land access, health and safety, permitting and compliance, SCADA systems, and engineering content. I find that oil and gas companies favor data diversity over narrow data standards, which typically means high numbers of systems, low levels of compatibility between systems, and high levels of manual integration and data mapping. Add in a tendency among asset-based businesses to separate commercial IT systems teams (looking after ERP) from operational technology (OT) teams (looking after SCADA) and the result is increased manual integration needs, not to mention limited insight into the cost of operations.

Legacy management approaches in production are now inconsistent with our increasingly digital world. Information about wells may be divided up among plays, which at one time made sense, when the cost of computing was high, but no more. The lack of data standards and production departments' practice of maximizing specific wells with bespoke analytics blocks the power of the crowd.

The Future of Production

The industry is at a critical point: digital innovation needs to become a significant investment focus. The collapse in oil prices in 2014 focused attention on costs and productivity like never before, and it increased the competitive tension between the providers of capital and the producers, the portfolios of assets, and even between individual assets. One senior vice president at a leading gas producer confided to me in 2016 that their hurdle rate for considering business changes that could reduce cost was at a low of zero for the first time in a generation. In other words, any idea that improved cost or productivity would get a solid hearing. In previous years, the hurdle rate for digital innovation had to exceed the hurdle rate for a new well (30 percent or better) and sizeable in absolute terms (better than $10 million), which would have sent most innovations to the sidelines.

I do not foresee a rush to digitize the existing (or brownfield) production assets. With so many to tackle, all of which are in decline and with an uncertain fossil fuel future, it will take many years to figure out which wells warrant investment and how much investment they can tolerate and still generate a return. With the average lifespan of a well measured in just a few years, I suspect many will struggle to build the business case to retrofit digital technologies given their productivity, location, costs to upgrade, and available infrastructure (specifically telecommunications). In seven to ten years, many older conventional wells will have declined sufficiently that shutting them down is the only plausible choice.

However, a real breakthrough could certainly come from applying digital innovation to the existing datasets associated with a producing asset. Better analytics of existing data, through machine learning, could expand the understanding of reservoir porosity and permeability, stimulation approaches, and fluid behaviors for the existing brownfield assets, which in turn could lead to better production decisions. For many older fields, with low-productivity wells, producers will be understandably reluctant to allocate much time and computer attention to revisiting legacy assets, but someone else might. It is estimated that only 40 percent of the known oil in conventional oil fields has been extracted, compared to 95 percent of the gas in conventional gas fields. Combining the data from multiple oil fields—by moving it to the cloud and subjecting it to improved analytics—could be the key to expanding oil recoveries to match those of gas.

In time, I expect all new wells to become progressively more digitally enabled. The rise of unconventional oil and gas wells in North America has forever changed the context within which such investments are made. These unconventional plays differ from their conventional peers in key ways that favor digitization.

First, the amount of hard capital in play (cement, steel, and equipment) is much less on a shale well than on a large offshore well, since most unconventional wells are much smaller in terms of scale and productivity and can be drilled more quickly and with more repeatability than conventional wells. The fast cycle times of these wells is better matched to the rapid pace of digital innovation. As a result, engineers should be able to design and embark on digitization programs secure in the fact that the well will be delivered quickly, with lessons learned and improvements captured in subsequent wells.

Second, the cost of digital has fallen and continues to fall over time, such that the amount of digital capital investment to be recovered is much lower than ever. Digital capabilities can make sense even on the most marginal of wells (those with relatively high cost and low productivity). I suspect that some marginal wells will be brought online simply because digital innovation enabled them to be successful.

Finally, unconventional basins can consist of hundreds of nearly identical wells. For an engineer, an unconventional resource play offers ideal conditions to introduce some new digital capabilities at each of its new wells. In other words, the learning curve possibilities that manufacturers enjoy through the repetitive delivery of produced wells can now be brought to bear on oil and gas assets.

Wells could move to an entirely new performance basis, beyond the supervisory and controlling systems capabilities of SCADA, and towards self-supervision and local agency (using blockchain to organize its own servicing). They could behave more like stationary robots and less like dumb machines controlled from afar.

Well-performance data, perhaps camouflaged for commercial reasons, could be moved to the cloud, where it could be aggregated with production data from other producers. Cloud computing and AI tools in the cloud could bring the most advanced analytics to the most junior players, whose understanding of their geology could be enhanced by the power of AI working with these larger composite datasets. That understanding could help solve the mysteries of the decline curve and enable producers to plan for optimal well intervention to keep production high. Well-performance data in the cloud could also unlock a new business model that separates the data about the asset from the asset owner and the service provided.

Cloud computing could also help improve service delivery in the field. With service providers, wells and assets, well operators, landowners, and regulators all working with the same data, services could be both optimized and improved. I go more into this idea in the section that follows on field services.

As I have described in the previous chapter, there are several other important illustrations of how digital will transform production. One is how IBM Watson is transforming engineering services in the offshore fields of Western Australia. Another is the use of blockchain to transform how royalties are calculated, which is a production function. And the final illustration is the use of aerial drones to supervise oil and gas assets from the air and submersibles to maintain production infrastructure deep underwater.

FIELD SERVICES: KABOOM!

Fuse length: one to three years
Impact: 4/5
Key technologies: collaboration, Internet of Things, blockchain, cloud computing

I am already seeing a wave of digital change in field services, driven first by the presence of so many smartphones in the pockets of workers, as an integral part of next-generation vehicles, and with the rise of smarter equipment. Many in the industry suggest that the area of field service is ripe for transformation along the lines of what Uber has done for the transportation sector.

The interface with a service company to deliver a service to the wells is, for the most part, still a manual exercise dating back to when labor was cheap, services were relatively inconsequential to the revenues from a producing well, and mobilization costs were irrelevant. The world has now changed, and the interface between producers and field service companies needs to adapt.

The Legacy Service World

Historically, the characteristics of the onshore field assets dictated a manual, human-centered field-service business model. In North America alone, the industry estimates the existence of at least a million oil and gas wells (no one quite knows as many wells are "lost"), and this volume grows by several thousand new wells each year. While the numbers are smaller, a similar situation exists in Australia and many other basins around the world. For the most part, wells will operate for many years before they eventually run out of production. Wide variances in well designs, flow lines, compression assets, and other infrastructure translate into the need for high skill levels in human workers and a broad range of services for those wells.

The asset owners tend to organize themselves along asset lines, which brings a strong focus on restoring assets to service as quickly as possible. The complexity of the assets, the range of potential services, and limitations in the service sector compel the asset owner to a manual, human-centered approach to service provision—finding a supplier, contracting the service, exchanging documentation about the asset, arranging permits and access. In theory, smaller, nimbler players offer a lower unit cost (or day rate) than large, integrated services players, but they can often struggle to innovate when under pricing pressure.

The offshore world differs in a few key ways from the onshore world: namely, there are far fewer wells, but the wells produce considerably more hydrocarbons and, as a result, the assets are more complex. Offshore services are usually concentrated among a small number of larger suppliers, with bigger assets (like fleets of service vessels and helicopters), substantial port facilities, bigger warehouses, and deeper crew depth to handle the more demanding subsurface needs as well as the more hazardous offshore platform services.

Suppliers are quick to point out the shortcomings of the legacy approach to field-service management. Labor is very costly now, relative to hydrocarbon prices. Mobilization costs are off the charts, and services are pricey. There's little to no coordination with and between the operators and their assets to optimize overall service. Business processes are overly reliant on lightweight tools (Excel, telephones) to coordinate service delivery. It doesn't scale well, which is a problem with the steady growth of onshore unconventional fields, which demand high levels of repetitive work.

Operators struggle to keep tight track of the performance of suppliers, so poor performance isn't corrected quickly enough. (Suppliers have lots of horror stories about showing up on-site only to have to demobilize because the previous service provider wasn't done yet.) The manual approach drives higher cost in that it doesn't allow for optimizing the match of services to assets needing services.

Contracting models are still traditional in comparison to other sectors: they do not include reverse auctions, swaps, and other clever contracting methods. Even building up the service order is a significant manual effort, which demands assembling all the required data about the well, the service need, the landowner, and relevant permits.

At heart, this is a coordination problem. Operators rightfully concentrate on keeping their assets running, whereas service providers rightfully seek to keep their kit and crew busy. In deep markets with lots of service providers close at hand, a concentration of conventional high-production wells and infrastructure, and strong commodity pricing, this isn't that much of a worry. An ERP or asset-management system kicks out a purchase order to some known servicer, engineers manually dispatch all the requisite diagrams and specs, and operations coordinates all the players using Excel or Primavera P6.

I challenge whether the services-coordination approach adopted from conventional oil and gas operations is still fit for purpose, if it's adequately leveraging available technology. The number of assets operating, their steady growth, diversity, long operating life, and the frequency of required servicing is at odds with workforce reductions and cost pressures in commodity markets, as well as ongoing regulatory pressures to reduce emissions. What would Toyota do if they were in the asset-service business?

Service companies have special advantages in the more digital future. To survive, service companies have had to innovate, develop new services offerings, invent tools, and adopt techniques into their business. This innovation gene is going to be put to good use. The cleverest field companies will take advantage of the distributed computers out there with better collaboration throughout the supply chain.

Collaboration across Suppliers

Many asset owners have already invested in collaboration systems with their preferred suppliers, but a transition is underway. The legacy systems that coordinated services to assets are mature and predate the

Internet age. These old systems, much like the ERP systems, did not anticipate the mobility age, the presence of smartphones in pockets, the arrival of smart devices on assets, and cloud computing. Over time, owners will refresh these systems with offerings from companies like ServiceNow, Salesforce, IronSight, Payload, and many others.

These collaboration systems will bring better geographic understanding of service delivery, as the entire relationship can be tagged with GPS coordinates. Driving routes will become visible, time on-site will become highly accurate, and handoffs from one service company to the next will become more efficient. Demurrage (stand-around time) will shrink. Service companies should see asset-utilization rates improve, and owners should see reductions in incidents, as well as their assets returned to operation faster.

The challenge is that suppliers will inevitably find themselves dealing with multiple systems in the run of the day as their crews switch from one customer to another, and from one collaboration system to another. I can foresee a need for service companies to only need one digital solution that translates between multiple customer systems. This could be a role for blockchain, for example.

Data Availability

Service companies will add more sensors to their own equipment to improve the quality of their services. Those sensors will collect data about the service being delivered, as well as data about the asset for which the service is rendered. Some may be able, and willing, to share both asset and service data in real time with the asset owner.

Few service companies actually use customer asset data to figure out how to create value through data. A downhole tool company will have lots of sensors to collect reams of data about the drilling activity, but who owns the data? Most service companies assume the data belongs to the asset owner, and few ask to keep a copy for their own use. I can see this changing as service companies conclude that the data has value. Some will soon make the case to the customer that they can assist with analysis of the data, in part by comparing data across multiple service calls.

The model for this is GE's turbine business. GE keeps tabs on all its turbines (including wind and gas) as part of its service arrangements with customers, by pulling turbine data into its own cloud computing and analytic environment. GE detects issues faster, brings predictive maintenance to the turbines, and moves fixes quickly throughout the fleet. This know-how will be quite valuable when all kinds of assets feature the same level of sensor penetration as turbines. Other companies quietly provide similar services for important equipment in plants, like pumps. Now any company can replicate this service because the underlying technologies are digital, freely available, and low-cost.

In keeping with the growing volume of field-service data, service companies at a minimum will need to boost their analysis capabilities. I can see service companies investing in upgraded analytics and artificial intelligence tools to deepen their understanding of all this data.

Autonomous Kit Integration

Service companies will adopt more autonomous technology over time. Shortages of qualified people in the industry make autonomous technology more necessary than merely desirable. This is already happening in the subsea with submersibles and in the large oil sands mines with flying drones and remote-controlled haulage vehicles, but it should also show up in highway trucking and haulage, sand handling, drilling equipment, services rigs, and pipeline installation. The large offshore platform companies are already experimenting with remote-controlled drill rigs.

A new market for operating and servicing autonomous technologies will emerge. China is the world's largest buyer of industrial robots for their manufacturing operations, but it is already reporting enormous shortages of robot operators. The robots can run 24/7 without rest, so each robot may need three shift workers to supervise it. As more autonomous technologies come to the industry and displace field labor, new higher-skilled jobs will emerge in operating robots.

Aftermarket Parts and Critical Spares

One of the banes of services companies is fixing equipment in the field when parts break or wear out. The rigs can't normally be sent from the field to the repair shop—that would be very costly. Instead, the parts are shipped to the field for installation. Some companies are starting to experiment with 3-D printing to create replacement parts in the field and save the logistics time. Every oil and gas field has its local, dedicated retailer that sells safety gear, duct tape, gloves, and other consumables (the so-called rope, dope, and soap outlet). These could become the utility 3-D print shops of the future.

At a workshop I attended in mid-2017, several large industrial companies set out their forecasts for the impact of digital on their business sectors:

- A global equipment company (pumps, controllers, and SCADA) projects a 30 percent improvement in asset utilization by applying digital techniques to the installed base of gear. That's a huge productivity gain.
- Another global equipment company forecasts that almost all oil and gas equipment will achieve greater than 90 percent availability by eliminating unplanned downtime through digital. Across a large enough business, that's like finding another big plant hidden among the others.
- A global power company forecasts that new digital technologies embedded in industrial equipment (gen sets, pumps, turbines) will yield between 7 to 15 percent reduction in energy demand.

PROCESSING AND REFINING: LOW AND SLOW

Fuse length: five to ten (or more) years
Impact: 1/5
Key technologies: Internet of Things, AI, machine learning

The refining and processing industry will lag behind the rest of the industry in digital adoption. It will take much longer to incorporate digital innovation in this very physical industry, and those innovations are likely to be of lesser impact than in other sectors. The fact is that the vast majority of processing and refining infrastructure predates the digital age, so there's a lot of catching up to do. On a global base of 94 million barrels per day of capacity in 2012, the refining sector has grown to 97.4 million barrels per day in 2016, or just 3.4 percent growth in five years. The world's top ten largest refineries were all commissioned well before 2012 and designed in the years prior. Much of the growth in the industry comes from de-bottlenecking or adding incremental processing capacity at existing plants, not from entirely new (or greenfield) refineries.

Nevertheless, brownfield processing assets cannot be simply ignored. There is considerable value at stake—a mature asset whose utilization rate or throughput can be raised by better decision-making drops cash directly to the bottom line. In 2012, the throughput rate of refineries globally was 76.6 million barrels per day, or 81.3 percent of available capacity. By 2016, throughput had grown to 80.5 million barrels per day, or 82.6 percent of available capacity. Each percent improvement in throughput is like building a brand-new world scale refinery of 800,000 barrels per day. At an average refining margin of $5/barrel, those barrels yield an incremental $1.46 billion to the refining sector.

Another way to consider the scale of the brownfield opportunity is in avoided capital expenditure. One of the newest refineries to be built, in Canada, has cost about $80,000 per refined barrel to construct. Other recent refineries are in the $60,000 range. To build an 800,000-barrel refinery would cost somewhere between $48 billion and $64 billion, which would need to last thirty years or more. Few Boards in oil and gas, facing an uncertain future for fossil fuels, will have the courage to sanction large, new oil refinery development. Clearly, one solid answer to grow the business is to squeeze more from the existing assets.

The adoption of digital technology in the processing and refining sector is all about dealing with these brownfield assets. As an asset-intense industry, oil and gas creates value by optimally selecting crude oil to refine, by running and maintaining assets (such as turbines, pumps, and processing units) as efficiently as possible, and by keeping overhead low. Assets last thirty to fifty years and retrofitting assets for a digital world is an entirely different challenge compared to building digital into a greenfield asset.

Digitizing Brownfield Assets

Unfortunately, brownfield assets do not lend themselves readily to the opportunity presented by digital tools. Their existing data networks may be wrongly designed, lack available conduit for wire line, or have too much electrical interference. The facilities may require hardened devices that do not spark. The plants are running 24/7 and there is rarely a big enough outage window within which to undertake a digital overhaul. Drilling holes in pumps to add sensors can't be done hastily. Even the people working at these facilities are brownfield, in that they have been carrying out their work in the same way using the same tools year in and year out.

The digital journey in brownfield should begin with an understanding of the operational performance of the assets in question. Closing any competitive performance gaps should become the target for digital investment. The Solomon benchmarks are by far the best way to understand refining asset performance relative to peers. Solomon is the most respected, complete, and comprehensive set of performance measures for plants, but the scope of the benchmark is limited to the refining and petrochemical sector. The benchmarks may not apply where management is actually not measured strictly on the economic performance of the assets they supervise.

A STATE-OWNED GAS operator in the Middle East wanted to undertake a right-sizing exercise a year into the commodity price downturn. While the company had been careful to complete the Solomon benchmark, in reality the company served a broader national interest by offering employment to the country's nationals. Employment levels were almost triple in comparison to the average of the industry. As a result, the benchmarks served little purpose in manning levels.

Benchmarks in processing typically point to three areas of focus for improvement. First are the potential gains from managing the overall asset—purchasing the optimal feedstock that best matches the plant configuration (as in the Goldilocks fairy tale—not too sour, not too light, not too acidic, but just right) and running the plant to yield the product slate that most closely matches the demand profile for finished products.

About 60 percent of the gap between theoretical value and actual value in refining stems from decisions made in managing the feedstock against the demand (and this is an enormous analytics problem).

Second are the improvement gains from plant operations, which include optimal energy usage, blending inputs to create optimal feedstocks, blending outputs to precisely match demand at the lowest cost, maximizing the use of expensive assets with spare capacity such as jetties and tanks, and improving the reliability of operating assets. About 30 percent of the theoretical value of refinery economics is bled away in these operational areas, and much of that value could be captured by better data and better decision-making.

Last are the improvement gains from decisions taken that do not directly involve hydrocarbons, including the level of critical spares inventories, people management, and minor repairs and maintenance. About 10 percent of the theoretical value of a refining operation is lost through poor decisions outside the actual hydrocarbon value chain.

TEXMARK, A HOUSTON-BASED chemical plant, kicked off its digital journey by taking a number of employees on a field trip to visit a digital lab hosted by HP. The employees were accustomed to using clipboards and paper worksheets to track critical asset performance. The lab demonstrated how Texmark, with limited investment, could add sensors to a handful of key pumps at the plant, which would send live performance data to tablets and give operators early visibility to pump performance. By predicting which pumps were approaching failure point, workers could schedule servicing and boost the plant reliability.

Often it is left to management to apply their judgment to identify gaps that they think should be closed. This approach has several shortcomings. Managers may not understand their current performance very well or their performance gaps relative to their peers. Their proposed targets may be either too aggressive or not aggressive enough. They may not select the biggest opportunities for improvement, or they may select change projects that are easy to achieve to assure a positive year-end review. Their agenda may not lead to innovation.

An alternative to management judgment in the absence of good quality benchmark data is the value-loss model. Oil and gas assets can often be modeled to reveal their theoretical or physical limits, and when compared to actual performance, the model can reveal the value loss or value at stake. The loss could become the target for business-improvement initiatives, some of which may be enabled by digital tools and techniques. The value-loss model removes some of the bias that might exist where managers simply set their own performance targets.

Of course, the gap in performance between pacesetter refineries and bottom-quartile performers is not explained by digital. After all, just about all refineries predate the wave of digital innovation of the past five years, and the shortage of investment capital has likely limited digital investments everywhere. Simply copying the best practices of the pacesetters could close the gaps in performance. However, much as how telecoms systems in China and Africa leapfrogged from wire line directly to wireless, digital could help laggards dramatically improve their performance, as well as help pacesetters accelerate.

Digitizing Greenfield Assets

With the physical, data, and communications constraints that create challenges in digitizing existing assets, one might think that digitizing a greenfield asset would be relatively simple. True, it is simpler than brownfield, but it still requires good planning to be successful. Smart owners will build in digital capabilities from the ground up, rather than design the facility and then add a thick layer of sensors on top. Thinking digitally from the start will let owners break long-held assumptions about the business and reimagine process flows, product receipt, inventories, maintenance, plant-monitoring, staffing levels, and supervision.

The value of digitizing a greenfield asset is that owners can use new capabilities (data, analytics, and connectivity) to radically transform the way business is done. What if a processing plant had the ability to almost instantly change energy inputs based on market pricing as processing takes place, to protect margins? What if it could instantly send test results of a custom chemical being tailored for a high-value customer? What if customers could monitor the manufacturing process directly to help validate the quality of the mix? To transform the business to achieve these goals, companies need to think differently, adopt a different longer-term direction, and road-map their way to achieve breakout performance.

Building a digitally enabled greenfield asset is also an excellent way to deploy an "Agile" team. (I'll talk more about that in chapter 4.) "Agile" thinking could enable the creation of radically different concept refineries—where technical, operational, safety, compliance, and business teams collaborate on the concept before the detailed drawings are commissioned. Creating the next-generation processing plant with digital built in demands the engagement of a cross-functional team early in the design process.

The Future of Refining

With long transition timelines to introduce change while technologies advance very rapidly, the refining sector will have a more challenging time getting in front of digital.

Aside from crude feedstock, the next biggest cost in a refinery is energy inputs. Small improvements in energy usage can drive big savings, and this is an area where digital could be applied. Energy use is a key component of industry benchmark data, but, as with any complex system, energy consumption in the moment varies from the average. These variances may not be picked up in the linear programs (LPs) and other models of the plant. Energy-use data feeds are crunchable by cloud-based AI tools. Better analytics might lead to very minor energy input changes that bring relatively little risk to plant operations and could be readily implemented.

FULL-PLANT OPTIMIZATION

Refineries are active users of modeling solutions. The refinery LP helps configure operational parameters to match demand for refined products and available crude oils. HYSYS and related technologies provide highly detailed operational models of a refinery. Managers and employees have their own rules of thumb and accumulated experience about the behavior of plants.

There is space for a different kind of digital version of the plant. A digital twin would incorporate variables that the LP does not address, such as the weather; would reflect assets that might not be correctly modeled, such as tanks; and would allow for more scenario generation and analysis, using cloud computing to drive the calculations. Cloud versions of the plant could overcome other limitations, such as different LPs for different parts of the plant or for merged refineries where management systems and data are still separated.

The latest LNG plants to be built are already incorporating a full, end-to-end business digital twin.

INTERNET OF THINGS

Turnarounds and shutdowns present the best times to introduce digital sensors and devices into plants. Each major asset in a large facility is eventually taken off line and repaired, not every year, but certainly every two to three years. Digitizing an entire plant in one turnaround event would be an enormous and costly undertaking, but pacing and staging digital introductions to match the turnaround schedule would slowly but surely transition the plant to a more digital state. Replacing a pump with its fully digital version might take a few years, but sensors can now be simply strapped onto an asset and capture much useful operational data, such as sound, heat, and vibration. Focusing on the most critical equipment in the plant could improve its overall resilience and reliability.

Before introducing sensors, managers should measure the baseline performance of assets so that sensors can be calibrated and capture actual pre-sensor performance. In tandem with introducing sensors, either to kit or as wearables for the employees, managers and supervisors will need new capabilities both to transmit sensor data (which may entail network investments) and to process the data that the sensors generate. Investment dollars should be set aside for dashboards, analytics, and machine learning.

This future state depends on changing the way the processing industry manages its data assets. Instead of siloed, department-specific data that is time-delayed and not available for decision-making, processing plants will need to rethink their approach to data.

FUEL RETAILING: THE FUSE IS LIT

Fuse length: five to fifteen years
Impact: 3/5
Key technologies: automation, Internet of Things, cloud computing, apps

Most of us engage with the petroleum industry at the consumption end of the value chain—fuel retail. At a meeting I attended in early 2017, an executive from a global oil and gas company posed the following question: What will be the impact of digital technologies on gas stations? What's to become of the thousands and thousands of gas stations in the future? His company sold petroleum fuels through a worldwide network of some 17,000 fuel stations, some of which his company owned outright and some of which were licensed resellers.

He posed this rhetorical question to the audience in light of several advancements that suggested that the retail fuel business model was ripe for change and, in some locations, already changing. Plainly, as electric cars don't consume fossil fuel, an eventual shift to electric cars could strand an enormous amount of fuel retailing infrastructure, including the full value chain set of assets such as the trucking fleet, tank farms, related pipelines, and ships. To get a sense of the economic risk facing fuel distributors, let's assume the average gasoline station carries a book value of $2 million (the land, building, and equipment). This is perhaps too low for developed countries where land may be much more valuable and too high in developing countries where vehicle penetration rates may be more modest. There are some 3 million fuel retail sites globally. This asset class is worth as much as $6 trillion. That's a lot of shareholder value that is slowly going to erode.

It's not just electric cars, however, that trigger shifts in fuel demand. Digital capabilities in gasoline-powered cars will have a similar impact.

The New Advanced Vehicles

It's comparatively easy to change the digital features of cars. Your next car, if it's a new model, will likely have considerable silicon in it in the form of onboard computers, sensors, LIDAR, radar, gauges, actuators, displays, and controls. It may be a hybrid model (with both a gasoline engine and electric motors and batteries) or it may be fully electric. It may run on hydrogen. It may be a node on a blockchain. Regardless, carmakers will stuff more digital smarts into their vehicles to help them reduce fuel consumption and connect their cars to the cloud.

ACCORDING TO EV WORLD, a conventional gasoline car in 2015 had up to fifty pounds of copper in it principally for wiring. Through digitization and electrification, fully electric cars in 2025 will contain 132 pounds of copper.

Our silicon cars are going to spew a progressively larger and more interesting torrent of digital exhaust in the form of data about engine performance, brakes, driving habits, fuel usage, music preferences, routes taken, and ridership. If we've learned one thing over the past decade, it's that data, no matter what it is, has value, especially large consumer datasets. We also know from experience with Facebook, Apple, Google, and Amazon that consumers will trade personal data in exchange for something of value.

There are three car-specific digital capabilities—autonomy, connectivity with each other, and shared usage—that will be progressively introduced into vehicles that may create downward pressure on fuel demand.

Once vehicles become autonomous (and they don't need to be fully electric to be more robotic—cruise control is a kind of robot), they can take over more of the driving. No more jackrabbit starts, hard braking, idling while waiting, and inefficient gearing on hills. Vehicles can be driven perfectly. This equates to fuel efficiency and less fuel demand. By how much is uncertain, but a recent case study by some Tesla drivers shows that a Model S could drive double the battery capacity of 300 miles (480 kilometers) on a single charge by driving for efficiency.

Once vehicles become connected to each other and to the cloud, they can talk to each other. Cars can then drive much closer together because the computers can take over the brakes (a computer's reaction time is 100 times faster than the best human reaction time). By driving closer together, cars can draft, just like bikes in a peloton in the Tour de France. Drafting is energy efficient—cyclists gain a 40 percent productivity improvement when they cycle in formation, geese get the same benefit flying in V formation, and so will vehicles on the highway. And cars don't need to be electric to be connected.

As vehicles become shareable, through services like Car2Go, Uber, Whim, and Lyft, digital eliminates the need to own a vehicle at all. Digital potentially reduces the demand for new and replacement vehicles. Most personal vehicles are used just a fraction of their available capacity, anyway—they sit idle in garages and parking lots for much of their lives. Assume a standard personal commuter car drives 12,000 miles (20,000 km) per year at an average speed of forty miles per hour (sixty km/h). That's a utilization rate of just 3 percent. To many millennials, it doesn't make sense to own something that expensive that gets so little use when a shared vehicle is around the corner. The shared vehicle might even be far superior in quality and comfort to what they could own personally.

Raising vehicle utilization rates from 3 to 6 percent, all other things equal, could reduce the number of vehicles we need to provide the same amount of transportation service by as much as 50 percent. Vehicle companies should be concerned should this business model of shared cars take off (and it is, judging by the interest in owning fleets of rental-on-demand cars). Fuel companies should equally be concerned about future fuel demand.

As the shared car fleets grow, fleet operators will increasingly eschew older gasoline models in favor of newer models that incorporate autonomous, connected, and electric features.

A COMPANY IN Sweden is experimenting with recharging vehicles as they drive. eRoad Arlanda has fitted a one-mile stretch of road with an electrified rail that will recharge batteries while cars are in motion. The system works with vehicles specially outfitted with a power coupling that lowers to come into contact with the rail. The system automatically bills the driver for the power charge. Other innovations, such as embedded induction charging systems, are in development.

These same capabilities (connected, autonomous, and shared) also apply to the trucking industry, which is even more susceptible to these technology developments (except perhaps the sharing phenomenon). Truck operators aim for high utilization already (trucks only make money when they're driving), and so their biggest levers to improved economics are to reduce fuel consumption (i.e., connected and automated) and increase load levels. The truck fleet also turns over much faster, and there's only 330 million trucks worldwide. Thanks to the reduction in coal-fired power generation, trucks are suddenly one of the biggest contributors to greenhouse gas emissions and the trucking fleet is growing. Truck operators will soon start to face the same emissions scrutiny as the coal industry.

The trajectory looks clear: over time, cars and trucks are going to become more digital and robotic, but not at the pace of the smartphone. Vehicles are simply more complex. But with dozens of truck and automakers out there, real pressure to reduce emissions, and no limits on the movement of good ideas and technology in our connected existence, it's not infeasible in this world of cloud computing, big data, GPS, and the Internet of Things to imagine vehicles eventually turning into clean robots for occasional or shared personal transportation.

Today, our cars have no idea which fuel it's been fed. Was it brand A with its emulsifiers and fuel markers or brand B with its engine-cleaning compound? At best, our cars might tell us how much fuel we consumed to go a set distance, but not what that cost us. And we really don't know our carbon contribution. We're supposed to piece together vehicle performance in our heads. Frankly, this is a problem for digital to solve.

Then again, your next car might not even be a petroleum car at all. Maybe it will be wholly electric, with replaceable batteries. It might recharge by parking over power plates in parking spaces. Or maybe it will plug itself into an electrical outlet, like a carpet-cleaning robot, in your garage or at a shopping mall. Or maybe it will run on hydrogen, compressed natural gas, or petroleum gas.

Certainly, some demand for petroleum fuel is going to permanently disappear, as many customers, particularly in younger generations, will switch away from it to do their bit for the environment. Other fuels may take its place, and that creates the need for customers to purchase that fuel—a possible ongoing role for the fuel station in the future.

Current State of Fuel Retailing

If there's a purchasing experience that hasn't materially changed in a very long time, it would have to be petroleum. We pull up to the pumps, get out of the car, program in what we want, pay in advance, grab the nozzle, dispense the product hopefully into the fuel tank (and not on the ground) and, occasionally, enter a booth to pay someone for the experience. The product smells bad and you wonder about the toxicity of the fumes. The odor of spills lingers for hours, and it's been this way for fifty years.

Sure, there's been some changes (getting rid of attendants, paying at the pump, the addition of a two-way radio for when you need help, the introduction of loyalty programs) but in the main, it hasn't changed. Consumers are simply doing more of the work.

Fuel retailers do not truly understand their customers the way Google does. Visa and Mastercard send you the bill if you purchase on credit, and your debit card doesn't transmit an address or any other personal data. Unless you're on a loyalty program, you're effectively anonymous.

Step inside a gasoline station and, aside from any obvious branding, they all look alike, configured the same way. The station agent is behind bars or wires, or in a glass cage (how's that for a job?) and typically near the door. In some countries, there may be a heavily armed guard at the entrance. Even the cash register has barely advanced in twenty years. The store layouts, with candy and junk food placed down low, are decidedly unfriendly to parents who would rather not leave their kids in the car while they go into pay.

The gap between the retail experience in the fuel world and groceries, apparel, footwear, gifts, music, books, and any other category is wide and getting wider. Leading retailers are rolling out dramatically different experiences to address customer needs, including cashier-less shopping, click and collect, and home delivery.

Aside from the arrival of electric cars, a second threat to petroleum retailing is the invention of "Mobility as a Service," or MaaS. Pioneered in Finland, MaaS offers its customers transportation services at a fixed price. Using a digital wallet, customers have unlimited access to shared cars like Cars2Go, bicycles from Ofo, subway, trains, buses, taxis, and hire services such as Uber and Lyft for a fixed monthly charge. In almost every city I've ever visited, these different transport services are distinct and siloed, requiring separate registration and payment setup. With MaaS, consumers may no longer need to own their personal transportation asset, which would probably be underused, anyway. The monthly fee for the service may well be less than the monthly insurance premium for a personal vehicle. The economics become very compelling, particularly for the next generation of consumers who are used to sharing assets and frequent short-term rentals.

As a side benefit, the service will generate an enormous treasure trove of consumer data that reveals buyer behavior presently hidden, including journeys, shopping patterns, underused routes, excess service capacity, and high-value services.

The Future of Fuel Retailing

If you've been paying attention to Tesla and its new automotive experience, and if you run a fuel retail network, you have reason to be a little nervous. The Tesla brand (fast, sexy, digitally advanced, self-learning, plug-in electric cars, giga battery factory, and a solar roof) potentially makes the retail fuel station an obsolete concept. Tesla opens up a world of possibilities for the consumer, and one logical implication is that Tesla owners (and probably all owners of electric vehicles) won't need to stop into a gas station for fuel. Could the shift to electric vehicles strand an enormous amount of oil and gas infrastructure?

I anticipate a very different future for fuel retailing, one that combines the digital exhaust from the vehicle, the transition in fuels, the shifts in retail technology, and the digital savvy of the next generation of drivers.

Imagine CloudCar, a vehicle app downloadable from the Apple App Store, as easy to use as an airline check-in app, where you set your vehicle and its drivers as fuel customers. You pay with your phone, so you don't need a dongle or loyalty card (yes, they're obsolete, too). In time, you might not even pay for the fuel as you dispense it—you pay as the volume in the tank goes down, a service enabled by blockchain. This would save needless visits to stations for those people who can't afford a full tank and only purchase an affordable amount (not to mention a clever way to really understand customer fuel-consumption habits).

CloudCar uploads driving data to the cloud and marries it to the app where you have your fuel purchase data. Bingo—now the loop is closed—the cost and carbon content of the fuel is linked to the digital exhaust of actual consumption and driving behavior. A digital version of your car, fuel, and emissions is in the cloud.

The app could provide useful insight: different drivers register via their phones that they are in the car and driving, which reveals who drives how much and how often, how much fuel they consume, and how much more emissions-costly some drivers are than others. Your teenage kids pay for the gas they use.

Through CloudCar data, city planners could calculate the carbon costs of highway bottlenecks as fuel usage, carbon emissions, and driving behaviors are linked to specific infrastructure. The carbon costs of city parking policies are clearer as CloudCar shows how much time and fuel is spent circling for a parking spot.

CloudCar could inform how your fuel carbon footprint compares with others in your neighborhood. Nothing drives behavior more than giving people real data about how their use of a resource compares to their peers. Driving is gamified around carbon targets, with drivers posting the lowest-carbon routes to work and getting "likes" from their followers.

NEW COMMERCIAL OPTIONS

High-quality consumption data will unlock more possibilities for creative fuel sale. For example, customers know that at times of high demand, fuel prices rise. Retailers, knowing how much fuel is in the tanks, could offer customers the option to purchase fuel ahead of times of high demand so as to smooth out lineups at the pumps, reduce volatility in deliveries, and eliminate run-outs and canning.

Gasoline retailers could begin to monitor fuel usage at precisely this inflection point in automotive technology change, when pressure to decarbonize steps up and alternative fuels take market share. This data could help continually tune fuel supply lines, from refinery runs to wholesale stock volumes to supply movements in the chain.

Gasoline delivery startups are already challenging the traditional retailing business model. These innovative digital services, enabled by cloud computing and apps on a smartphone, are comparable in price to traditional gas stations. Rather than stopping into a retail station on the way home from work, consumers can arrange, using WeFuel or Filld, for fuel to be delivered to their vehicle while they are working or shopping. Imagine driving your car to the mall and CloudCar, detecting a fuel need, requests a tank top-up delivered right to the vehicle, with the purchase recorded and settled using blockchain. And while that service could carry a premium charge, fuel delivery to the car frees up valuable time for the consumer. Fuel delivery to the car threatens the relevance of retail fuel stations.

Fuel retailers are already experimenting with owning the fuel tank in the car. Companies with interests in energy such as Eni S.p.A., Bolloré, and Repsol are investing in car-sharing fleets like Enjoy, Autolib, Blue-Indy, and Ibilek; in a future of self-driving vehicles, these cars would always return to their branded gas stations to fuel up.

The next innovation in a new fuel retail experience could be robotic refueling, or FuelBot. The advances in optics processing, autonomous technology, machine learning, and cloud computing mean that a single flexible robot should be able to figure out over time how to refuel a specific make and model of car (where the fuel cap is, how to open the fuel cap door) and then transmit that learned experience to every other similar robot at the same time. Fleet learning comes to fuel retailing. FuelBot could be particularly economic when all those unmanned self-driving cars trundle off in the night to refuel. And if FuelBot and CloudCar talk to one another, FuelBot could make a good educated guess, based on driving habits and measured fuel in the tank, how much fuel to dispense to the physical car and suggest that amount to the customer.

Aside from CloudCar on the road and FuelBot at the pump, what could the experience be like at a gas station of the future? What could the gas station become? The station of the future could be a lot smarter, more like a community energy center, with solar panels on the canopy, rooftop, and embedded in the pavement around the station. It could supply gasoline for robot petroleum cars, plug-in power and replacement batteries for EVs, charging pads for contact-less power-charging, CNG for natural gas cars, and hydrogen for fuel-cell vehicles. With robot cars visiting at all hours to refuel, the gas station of the future could be more highly utilized and could even take up less room.

A REVAMPED CUSTOMER EXPERIENCE

The self-checkout stations I love in grocery retailing could soon come to gas stations. As retailing moves more definitively to cashless operations, with more vending machines and CloudCar–enabling fuel purchases by smartphone, a SmartStation of the future would lessen the issue of armed robbery, with no cash on hand.

CloudCar could also help direct customers into local SmartStations. Indeed, customers are already expecting a seamless experience across these various channels. Look to the Asian convenience retailer, models like Walmart's Pickup & Fuel, or even Starbucks' preorder mobile app. CloudCar could enable the pre-arrival order or even provide more precise arrival times for drink prep.

CloudCar could alert SmartStation that a preferential vehicle is approaching, premake a food and beverage order, or offer a SmartStation promotion based on previous purchase history. SmartStation could learn over time the impacts of weather, local events, and holidays and adjust shelf stock levels so that the most in-demand items are on hand.

The big convenience store chains are constantly introducing new products as product novelty drives traffic. SmartStation and CloudCar, working together, could push notifications to consumers about new items. Data will help target these notifications to just the right demographic.

A Long Transition?

Simple math suggests that it could take a very long time for the global car fleet as we know it to fully adopt digital features. There are about 1 billion cars on the planet in 2018, and carmakers churn out 72 million new cars a year with the current factories and supply chains. Of those new cars, about half are for replacement (as we trade up) and half are for growth in the market (the U.S., for example, adds 15 million new drivers every year). Some of the cars are to replace the ones we have wrecked in our too-frequent accidents, but a lot end up on the used market where they experience an extended life.

Even if we wanted every new car to be fully robotic or electric, to turn over the existing fleet of 1 billion cars it would take a minimum of thirty-three years with our current manufacturing capacity. A government program to buy out all the gas cars would take billions of dollars, which is a remote possibility with all the pressures on governments today.

While it's a good bet that people will have to drive themselves in a petroleum car for a few years yet, oil markets are rather sensitive to minor changes in supply and demand. It won't take much for electric and robotic vehicles to start to exert influence on global oil markets. Other parts of the world are going to get there first. Take China as an example.

The IEA reports that the worldwide fleet of electric cars reached 2 million in 2016. That isn't much, but it took twenty years for the first million and only a year for the second, with most of that demand growth in China. The Western world doesn't perceive this change in the market because it's happening in a distant place, involving low-profile companies whose products are not exported or advertised.

With vehicle ownership in China much below that of North America, first-time car buyers are more likely to buy the latest automated and electric cars. Driver expectations and behaviors will be set by the unique features of these new transportation solutions, whereas Western expectations for vehicles have been set by the incumbents (petroleum cars). Chinese consumers appear ready to embrace low-emission vehicles, which will help improve domestic air quality.

Once enough electric transportation options arrive in showrooms, customers will decide for themselves, of course. The top six major car companies account for more than 50 percent of vehicle sales, and all claim to be developing new drivetrains that rely on electric power partially (the hybrids) or fully (the all electrics). Ford, for example, clearly states that all electric is the way to go to achieve a zero-emission future. Hybrids may be just a stepping stone. Automakers are already advertising how much their new electric drivetrains will save consumers on fuel bills.

When will there be enough demand destruction of transportation fuel (particularly gasoline) that the shortfall will impact the price of crude oil? This is a twisty problem if the goal is precision. A blunt way would be to figure out when oil markets react to small shifts in supply or demand and how much gasoline a typical car uses each year. If you estimate how quickly cars can adopt digital and electric features, you could identify the point when digitally destroyed demand will trigger a price move.

Oil markets tipped into oversupply in June 2014 by a piddling amount; OPEC estimated anywhere between 1 to 2 million barrels of daily production. Markets are finely balanced, as 1 to 2 million barrels per day was about 1 to 2 percent of daily production (back then, somewhere around 95 million bbls/day). So, how many of these new digital cars does it take to displace 1 million barrels of crude oil per day? Let's call this the demand destruction tipping point.

The world consumes about 100 million barrels of oil per day, or 36.5 billion barrels per year. About 25 percent of crude oil (more or less) is converted to gasoline for gasoline cars, or about 9 billion barrels of gasoline per year. The billion cars out there must consume those 9 billion barrels of gasoline, which is about 9 barrels of gasoline per car per year. That's a really blunt-instrument approach to the numbers, and very imprecise, but it gives us a rough idea of demand per car.

A million barrels of crude oil per day, refined to 250,000 barrels of gasoline per day, or 91 million barrels of gasoline per year, supplies about 10 million cars per year at a rate of 9 barrels per car. Once we have 40 million all-electric cars on the road, we will have permanently destroyed demand for 1 million barrels of crude oil per day. The first 2 million all-electric are on the road now, and it won't take ten years to get to 40 million because of the cumulative effect of adding more electric and digital cars every year. A large market like China could achieve this volume much more quickly.

Of course, the demand for internal combustion engines in the short term will continue to grow, and with it, the demand for gasoline. But there will come a point where the accumulated base of electric vehicles, with a steep growth curve, overtakes the demand growth for petroleum, tips the market into oversupply, and causes chaos throughout the oil market value chain. Demand destruction will accelerate, and diesel demand destruction won't be far behind. What is more worrying for oil companies, and harder to predict, are the impacts of fuel-efficient, connected, autonomous, and shared vehicles, as well as the rise of electric scooters and bikes, which will alter demand even faster.

The leading oil companies are already anticipating this challenge and becoming warier of oil and gas capital projects that extend beyond 2030 or so. It is at this point when many models of petroleum consumption suggest we will be at the tipping point, where demand for oil in transportation starts to fall.

CAPITAL PROJECTS: THE LONG, LONG FUSE

Fuse length: ten to twenty years
Impact: 4/5
Key technologies: collaboration, cloud computing, blockchain

The oil and gas industry depends on the construction sector to build new infrastructure (oil refineries, pipelines, gas plants, LNG facilities, chemical plants, and tank farms) and to do so in an efficient manner. When oil prices are very high, or margins are robust, the costs of construction may not factor too highly into whether or not a project proceeds. After all, the costs of construction are amortized over the life of the project, which may be forty years or longer—an extra dollar here and there doesn't amount to much of a worry. A large spend program (say, $10 billion for an LNG plant) could be amortized to the tune of $500 million per year, a pittance to a business that might generate $5 billion per year in cash flow. As a result, asset owners have not been overly stressed about the performance of the construction sector in the delivery of new assets.

However, with the decline in oil prices, the industry has cut back dramatically in capital spending. From 2015 to 2017, the oil and gas industry canceled or postponed about a trillion dollars' worth of global investment, according to a study by the Deloitte Center for Energy Solutions. That investment would have been spent to prolong the life of existing assets and to build new assets. As wells decline in performance, to the tune of about 6 percent per year, the industry must invest capital to replace that production (or about 5 to 6 million bbls per day) and to fund for growth (another 1 to 2 million bbls per day). Eventually, oil and gas has to get back to spending capital.

The construction industry, however, is among the least digital of all industry sectors globally, according to McKinsey, which has studied the digital readiness of various sectors. Construction productivity in oil and gas worldwide has actually declined by 25 percent over the past two decades, whereas most other industries have advanced or become more productive. Digital innovation in the construction sector could have a very big impact on the trillion dollars of oil and gas projects unable to secure capital. As capital costs fall, the economics of capital spend improve, which encourages Boards to sanction new projects.

To illustrate, I was involved in a study to benchmark the Australian LNG projects in 2016 to assess their global competitiveness and to provide guidance on what changes the country could undertake to improve the fortunes of the LNG sector. The nation was anticipating that the demand for natural gas in Asian markets would continue to grow, the country would have some ten world-class LNG projects ready to expand, and the long list of stalled projects could potentially be relaunched to meet that demand. LNG is now the world's second-most valuable commodity, and a long list of countries (the U.S., Canada, Mozambique, Qatar, and Russia) are lining up to supply the hungry markets in Asia. Cheap and plentiful gas is certainly important, but so is the cost-effective delivery of the capital.

The Australians' track record in delivering these large capital works is not good. Most of its recent LNG projects were delivered either late, over budget, or both, and the government's expected royalty windfall has fallen well short of plan. The benchmark attempted to pinpoint where the projects were not competitive and what the nation could do to improve its position as a cost-competitive global supplier. The benchmark reviewed all known operating LNG projects and ranked them by break-even cost. Not surprisingly, the Australian projects were almost all on the highest end of the cost scale.

A key finding from the benchmark was that a 9-million-ton per annum LNG project could reduce its break-even cost by $1 per million British thermal units (or MMBtu) per year for a full twenty years by shortening the four-year construction project to three years. To put that in perspective, 1 million metric tons of LNG is 52 trillion British thermal units, so a 9-million-ton plant delivered twelve months faster creates a pricing advantage worth $468 million per year for twenty years.

The Builder's Standard Playbook

The oil and gas industry has a number of tactical ways to manage the timelines for the construction of its LNG facilities, starting back with the Qatar developments in the 1990s, which were fine-tuned in Australia and are now coming to North America. It begins with engineering design. The engineering world is now highly digital, with designs completed almost entirely digitally. Standards may not be universal, but at least the content produced is digital. Digital content can be reproduced more easily than the hand-drawn versions of yore.

Next is modularized construction. Instead of building assets on-site, they are built as a series of modules in overseas fabrication yards and floated to the site, where they are lifted into place and welded together. Multiple yards can build multiple modules in parallel. Certain items with long lead times, like steel, are ordered early. Of course, there is some commodity risk if the quantity of steel is not yet fully scoped, and buyers may order too much. Some expensive rental equipment (cranes, ships) that are in high demand also get booked for specific times on the schedule.

To keep labor costs contained, owners try to put in place union agreements that will last throughout the project's duration and negotiate fixed-price contracts with key suppliers. This pushes cost and schedule risks to contractors who are then motivated to remove from their bids anything that could be seen as adding to cost, without a guarantee of a return (like digital innovation).

Frankly, these are mostly contracting strategies, and while they do matter, they tend to dominate the cost and productivity agendas for building new facilities.

To test how much receptivity there is to digital innovation in construction, I met with two large engineering, procurement, and construction (EPC) firms to ask about how they react to any digital innovation imposed by the owners. The answer was "we raise our prices by at least 20 percent." The reason is that the EPC industry has invested in its own systems and methods for its purposes, which they are loath to strand, and which can be at odds with the owners' needs.

While incumbent EPC firms are in a strong position to push digital innovation onto the construction sector, they don't. Some simply aren't that mature themselves. Others are based in countries that discourage digital investment because of concerns about job loss. Meanwhile, the tools that could enable the construction industry to improve its productivity are available today, based on the same building blocks for all other industries (cloud computing, mobile devices, analytics, telecoms networks, cheap sensors, and robots).

The Future of Construction

A cloud version of plant design—not just the workings of specific assets, but the whole plant—is an immensely important asset. It serves as the basis for quality and technical reviews, workforce mobilization, and stakeholder engagement. Reviews that surface design problems eliminate rework. Orienting and mobilizing the workforce faster also cuts down on delays. Showing stakeholders what the plant will actually do removes resistance to the project. A digital design can be fed into an augmented reality engine to give users a far more visceral experience and engage with the design.

Designs need not begin as a digital construct. Some industry leaders use sensors, drones, and robots to survey assets already built or in the process of being built to craft the digital version. The resulting 2-D and 3-D drawings of the asset will not be absolutely perfect, but they will be good enough to provide value to the builders.

A GAS COMPANY halted the development of a plant, which was being supplied by a fabricator, and left the entire plant disassembled in a large set of shipping containers. Eventually, the documentation of what was in the containers vanished, leaving the company with a perfectly good plant but no way to assemble it. A drone company, working with GE Predix, sent a robot with sensors into each container to build an image of what was inside (pipes, pumps, fans, valves, etc.), create an inventory of contents, and then configure the build instructions digitally, based on the container contents.

Why stop with the design? A solid design without a solid plan will likely take longer to execute. Today, industry leaders convert the construction plan to a kind of computerized video, which they subject to the same kind of quality and technical reviews, mobilization, and stakeholder sessions as the plant design. As much as 10 percent of the construction time can be cut out of the plan once engineers more easily visualize how the plan works in practice. The digital plan could eventually incorporate every worker, every rental asset, every tool, every permit, and every movement. This kind of big data problem has been crushed in many other industries.

With the plan and design now linked up digitally, artificial intelligence can be put to work to analyze better ways to both build and execute. The plan and its design could be tested and run millions of times under any variety of specific goals and scenarios to see its behavior and how it could be improved. Engineering logic might suggest that the least costly and most efficient way to build a facility could be by optimizing labor cost, but an AI review might suggest that optimization of shipping corridors or equipment rental could be superior.

A big issue in construction is that subtle design and build-interface errors are caught far too late. An interface is where one fabricated asset such as a pipe has to line up with another asset such as a pump. Industry leaders use LIDAR sensors to build digital versions of both the pipe and the pump at the point of interface to find those interface errors quickly. In this way, actual delivered assets match each other once they are mated, and construction delays are reduced.

Another issue in construction is that the scale of works can outstrip human capacity to understand and manage. Leading innovators are now using drones to fly over large construction sites to gather detailed data from the air, creating a digital version of work in progress that is used to inform the state of the build. Better data about current progress is vital to make better choices to keep a project to schedule.

A leading cause of execution delays is the creation and handling of the paperwork required to move physical things across various jurisdictions. Customs forms, inspections, manifests, packing lists, and tax calculations are the perfect use case for blockchain, which can remove delays in moving items through a supply chain by making data common and shared. Maersk has embarked on a project to revamp shipping documentation using blockchain, and it's estimated by the World Economic Forum that efforts to improve the efficiency of trade could help add as much as 5 percent to global GDP.

Construction sites can be a bit chaotic and finding necessary things, such as tools, jigs, and rental equipment that have gone walkabout, is a constant problem. Not so in a digital world. Inexpensive sensors can be affixed to just about any mobile asset or thing, including tools, equipment, and vehicles to give real-time visibility. It should be unacceptable in an era where we can track delivery of a $40 package to its precise location and arrival time that we can't see where the $2,000-per-day crane is on a construction site.

Studies repeatedly show that field-worker productivity in oil and gas is at best 50 percent because of late starts, early quits, missing permits, missing tools, and wrong crew skills, among other things. However, virtually every worker on a construction site now has a supercomputer in their pocket. These could be used to provide training, capture safety concerns, send and receive work instructions, issue permits, track tool usage, record hours, capture productivity data, and identify quality problems. Even the simple act of getting onto a site could be accelerated by using an app and a reader at the gate to let workers swipe in quickly.

The prize for construction is huge, but the barriers to adopting digital innovations are daunting.

TURNAROUNDS: SLOWLY BUT SURELY

Fuse length: five to ten years
Impact: 2/5
Key technologies: cloud computing, Internet of Things, ERP, artificial intelligence

Turnarounds and shutdowns are the practice of slowly and carefully shutting down a continuous operation like oil and gas production, carrying out a range of repairs and mechanical adjustments, and then slowly bringing the equipment back into service. Almost all oil and gas assets have some kind of scheduled turnaround process. Ideally, the time gap between turnarounds and shutdowns is as long as possible, and the time spent carrying out the work is as short as possible.

The impact of digital on turnarounds could be dramatic, since any reduction in a turnaround schedule hits several key metrics at the same time. A shorter turnaround is lower-cost, asset usage improves as assets are put back into service quicker, and revenues rise for every day that assets are productive.

Today's Turnaround

Recently, I met with a senior planner for an oil and gas outfit whose role was to improve turnarounds at his company. This is a complex undertaking; the range of repairs and adjustments that are included in the typical turnaround can be overwhelming. Everything from fixing a broken handrail all the way to recycling catalyst in a processing unit is carefully planned out and scheduled during the year (or the years) ahead of the turnaround event in a plan that spans many thousands of steps. On turnaround day, leagues of contingent workers descend on the plant and await instructions to carry out their assigned tasks. Parts and equipment arrive in a steady stream to massive lay-down yards ahead of use. A fleet of rental equipment (power units, cranes, hoists) stands by awaiting job assignment. Scaffolding and temporary structures are quickly erected to enable work at height.

Plants are all different, preventing the replication of good turnaround practices from one plant to the next. Turnarounds are also infrequent affairs, perhaps an annual undertaking, and institutional memory of what worked and what didn't fades particularly during times of high staff turnover. It can be hard to justify investing in improvements to turnarounds, when a turnaround might only last fifteen to twenty days.

Management, being mostly rational, recognizes that plants only make money when they're running, and so the goal is to minimize the amount of time and cost it takes to carry out the turnaround. Managers are motivated to double- and triple-order the number of workers and pieces of rental kit to eliminate delays in getting work started and completed. Building a big on-site inventory of these resources is a costly way to do business when margins are tight, particularly where turnarounds are in remote, hard-to-reach places like Canada's Far North oil sands or the North Sea offshore.

The costs of just about everything in a turnaround (people, time, regulatory compliance, equipment, emissions, energy, and consumables) have gone up, whereas the cost of digital has fallen to near zero. With so many computers on-site now, how might digital solutions be applied to the challenges of executing turnarounds?

It begins with having a solid understanding of the plant. If tags don't match equipment, if equipment records are out of date, if diagrams don't reflect actual installed equipment, then the plan is based on poor-quality data. It will be wrong from the very beginning. The biggest payoff from digital on a turnaround is to fix the data before the turnaround even starts.

The management tools and methods typically used for running turnarounds assume that the turnaround is like a project and not a process. Turnarounds show the kinds of cost and productivity improvements typical of projects, and not the more dramatic and impactful gains from the process-centric world of manufacturers. A manufacturer designs, implements, and refines the manufacturing process (say, an assembly line) to constantly remove waste and reduce cost. Turnarounds do feature some level of repetition (such as gate admissions and crew mobilization) that could leverage the kinds of tools we see in manufacturing.

Complex plans built by humans will inevitably have errors in them. Just as with construction projects, turnaround plans can be so large and complex that humans can no longer fully understand them. Expensive rental items can sit idle, racking up charges, because of hard-to-notice but avoidable errors in the plan.

Once a turnaround is set in motion, simply finding people, equipment, and tools on-site can be a nightmare. Many a turnaround features urgent text messages offering bounties to the first person able to locate a missing tool (although perversely this also encourages tools to go walkabout so as to trigger the bounty payment).

A large contingent and inexperienced workforce creates its own challenges. They need quick training on task and above-average levels of communication to address unforeseen issues in the moment. Digital could accelerate their productivity.

The Turnaround of the Future

Data is the great enabler in turnarounds. High-quality data unlocks many benefits, just as poor-quality data hampers even legacy manual work. It is painful work, but it is imperative that redundant, obsolete, and trivial engineering content be trimmed away, that equipment tags match the records, that diagrams properly link to the assets in the asset register. Many oil and gas companies have multiple incompatible systems that deal with engineering content. Long-life assets even have paper-based records on which the business relies. Leading practice is to capture asset data from the capital projects that created the data in the first place and load it directly and accurately into the asset register.

IMAGINE OPENING A recipe book and discovering the ingredients don't match the instructions, or the ingredients are correct, but quantities are wrong, or the instructions are out of order, or the temperature or time measurements are wrong, or the pictures don't match the outcome. Data-quality issues have no place in the kitchen.

Market leaders build a digital version of their assets, along with the tools, equipment, and people that will work together on the turnaround. This digital twin enables engineers to visualize the assets in the turnaround and help them understand space constraints, bottlenecks, and logistical issues. The digital version of the assets can be represented graphically on a computer monitor or, in more advanced settings, as a 3-D hologram that lets users virtually fly over, around, and through as they carry out turnaround planning.

An elaborate digital twin would also include rental equipment, scaffolding, spare parts, vehicles, inventory yards, and crews, along with their productivity measures, quality indicators, and costs. The amount of data in this digitized version of the assets would overwhelm human users but is well within range of modern computer systems.

Engineers who lead turnaround planning have perfected the use of Primavera P6 for this task. The pace and staging of activities, the duration of work, number of resources needed, costs, and equipment and tool needs are neatly captured in the tool. It records job and work actuals to provide deep insight into the execution of turnarounds.

Leading companies take P6 one step further by loading the plan into a software game engine and marrying it to the digital twin representation of the assets. Now engineers can "play" with the plan to see how it interacts with the assets: to surface work-sequencing issues, such as cranes arriving on-site to find no beams to lift; physical constraints, such as air drafts, turning radius, or bridge weight loads; productivity hurdles, such as queues and bottlenecks, inadequate parking, and insufficient lay-down space; and excessive investments, such as rental assets that never get used or underutilized crews.

Next, the game is "played" under various conditions, including foul weather, mishaps, shortages, and quality issues, creating enough additional data and insight that artificial intelligence engines can take over to work the plan to remove wasted time and excessive cost. After millions of iterations of possible scenarios, much like how AI learns to play chess, AI is able to help mitigate weather scenarios, the impacts of incidents, and other hard to anticipate challenges. AI helps uncover deeper plan issues and constraints and identifies the optimal use of people and resources in the moment. Engineers then apply more humanistic insight to get to an even better plan.

A COMPANY IN Australia, Real Serious Games (RSG) offers gamification services to heavy-asset industries, like water, power, and oil and gas infrastructure. RSG specializes in building game engines, the underlying technology for video games, which they apply to industrial applications like construction projects, turnarounds, and safety training. Working with capital project engineers, they build visualizations of how capital projects will behave according to the plan, and after the project they forensically study cost overages.

During the actual turnaround, the resources at work (people, tools, assets, equipment, consumables, and inventories) would need access to a network to be visible. Large sites will need to stretch a telecoms network over the area and leverage all those in-pocket supercomputers to enable visibility. The network could feed all the actuals back to the planners who will use AI and its scenario ability to constantly course-correct.

Many of the inanimate objects used in the turnaround (tools or rental equipment) can already be equipped with light-duty beacons to broadcast their presence. As soon as one of the supercomputers comes within range of a beacon, it can handle the heavy lifting of broadcasting the location of the beacon to the cloud. Planners should assume that eventually most, if not all, site assets will become active participants in the turnaround.

Turnaround planners and managers need to enable the field teams to execute the turnaround. This is challenging, because most of the turnaround workforce is typically contingent (that is, they are brought on board only for the duration of the turnaround). The demands on this team are high: they need to be in place when required; transported to the site; certified for the work to be carried out; equipped with the right tools; and trained to meet the safety standard. Planners are motivated to bring this team on board as late as possible so as to keep costs to a minimum.

Readying this team is a good role for virtual and augmented reality, or VR and AR. The workforce can be equipped with viewers ranging from Google's 3-D Cardboard viewer all the way to Microsoft's HoloLens, along with a set of apps that presents the digital twin and their activities in the plan for the day or week. Project engineers could annotate the visuals with alerts, cautions, task clarifications, and lessons from other turnarounds.

Field teams will often need access to some expertise in the moment—guidance to work with an unfamiliar pump or clarification of a diagram that doesn't match the actual field asset. AR tools, such as from Fieldbit, enable an individual or team to communicate with a senior expert back in a planning room. Camera-equipped eyewear sends the visual from the field to the senior engineer, who annotates the visual and that is shared with the field-worker in real time. AR keeps the team moving and avoids having a senior engineer waste time traveling on a big site.

Turnaround managers need to be able to communicate to the field in the moment with alerts, broadcast announcements, or specific task changes. Modern team collaboration tools like Slack are useful for this job. They are very easy to use, require little to no training, and already exist in mobile form for deployment on smartphones and tablets. Hundreds of apps and extensions have been built by creative users, including integration with calendars, time capture, documentation readers, and cameras.

While engineers need to see the whole P6 plan, workers and teams only need to see their specific tasks. Fortunately, there are several enterprise digital solutions for managing daily work and services, from companies like Salesforce and ServiceNow. These solutions also have readily available apps for smartphones and integrate with the technologies that underpin the digital twin, enabling delivery of diagrams, specs, instructions, bill of materials, and all the other key data directly to the worker. Technologies from companies like RedEye also allow for markup of diagrams, capture of work records, and time-tracking. These existing digital advancements could be applied to turnarounds.

Some jobs were never meant for people, and the oil and gas industry goes to considerable length to protect its people from harm. Eventually robots will be used for hazardous jobs, such as inspections of tanks—inventors have developed swimming robots that can dive into a full tank and check for repair needs; inspections at height given the predominance of flying drones; and in pipelines through intelligent inspection devices called smart pigs. These jobs will be controlled by humans, but the robots will be guided by the digital twin version of the as-built asset on the cloud. Autonomous wheeled vehicles will ferry parts and equipment to the right locations, around the clock, relieving the need for drivers and crane operators.

Digital tools might not change the cultural challenges of turnarounds: field rivalry; poor communications between home office and the field; organization models that block knowledge sharing and promulgation of lessons; and NIMBYism. But enabling transparency and better communications between field-workers and turnaround managers through digital advancements can make a difference.

SUPPORT FUNCTIONS: SHORT FUSE, BIG BANG

Fuse length: five to ten years
Impact: 4/5
Key technologies: ERP, Internet of Things, blockchain, cloud computing

At the end of our journey through the hydrocarbon value chain, we come to the support functions of HR, finance, and supply chain. I may be biased, but in my experience, these functional groups have their business needs largely met by one or several ERP systems, report writers, and analytic tools. However, the enterprise software fuse has been lit—the big ERP retrofit for a digital age is underway, and these functions have not much option but to respond. The bang is going to be very big, both in cost and impact on the business. The latest ERP versions are very different, the scale of change is massive, everyone is upgrading at once (leading to implementer unit-cost escalation), and other digital technologies will impact the functions at the same time.

For the commercial functions, such as finance, supply chain, and human capital, digital technologies may be adopted much more quickly. There are no energized assets that need to be carefully managed, and human work processes can be altered much more quickly than a manufacturing process. For example, robotic process automation (RPA) can remove a lot of the drudgery from existing work practices, like contract reviews, reporting, and invoice preparation. Blockchain technology will overhaul payments, field services, trading, land transactions, and supply chains. New ways to think about the supply chain will be opened up as 3-D printing allows for certain parts to be made on-site. But ERP solutions set the pace.

The Current State

Oil and gas companies, led by the supermajors and the majors, adopted ERP systems, particularly SAP, as a solution to cost and productivity pressures in the mid to late 1990s. At the time, commodity prices fell to as low as $10 per barrel, and opportunities for cost reduction and productivity improvements had been exhausted. Many other industries at the time (manufacturing, logistics, resources, public sector, consumer goods, retailing, and financial services), helped along by the systems integrators, adopted ERP technologies to replace dozens of incompatible bespoke solutions, unmanageable reporting, expensive mainframe systems, multiple databases, and a hodgepodge of system interfaces and Excel spreadsheets.

Over the past fifteen years, most big oil and gas outfits have found the lure of ERP tantalizing. These solutions are hard to resist when the industry leaders all use them, and Boards question your judgment if the business is inconsistent with its peer group. Today, the largest companies are now highly reliant on ERP systems to handle finances, sales and billings, purchases and payments, talent, payroll, reporting, inventory, supply chain, and assets.

One CIO compared the experience of licensing ERP technology to a home garage. Every few years, you can't find something in the garage (data problems), or the car no longer fits because there's so much accumulated stuff (an acquisition), or the kids move out, leaving their stuff behind (a divestiture). You set aside many hours (an ERP upgrade) to work through the detritus (process change), straighten out the shelves (reorganize), clean out the trash (eliminate old systems), perhaps run a garage sale (never). It might even be a touch emotional (change resistance), because of sentimental ties certain family members hold towards some art from kindergarten (still-working legacy practice).

Once it's done, you yell at the family members (change management) about how you want the garage to be kept tidy (sustainment team). Then you ignore the whole thing for a few years until it reverts to a fresh state of disarray (new release).

Hence the problem: not only is the garage messy but *all* garages are now messy. All industries have to change, and there are no alternatives to ERP. The last time this happened, at the turn of the century in response to Y2K issues, costs ballooned, and the ERP deployments ultimately did not achieve their fullest potential. The first implementations and later upgrades may not have taken full advantage of ERP's available features when implementation teams made arbitrary scope choices. Sometimes users could not bring themselves to change how they conducted business to match the capabilities of the ERP solution, and instead elected to modify the ERP system to match their process. Finally, governance structures (the committees and sponsors who make the decisions on IT investments) may not have vigorously enforced a clear preference to exploit their ERP investment.

Today, the support functions find themselves burdened with a costly mix of business practices that barely hold together, such as:

- a continued reliance on legacy approaches to doing business that predate the Internet age;
- a heavily modified ERP solution that is costly to maintain and upgrade;
- hundreds or thousands of additional solutions beyond the ERP suite;
- multiple ERP solutions in the largest oil and gas companies;
- lots of piecemeal integration software that tries to keep databases synchronized, common data consistent, and interconnected processes aligned; and
- a pervasive use of Excel to help glue processes and systems together.

During the all-too-frequent mergers and acquisitions in the industry, the costs to rationalize all this investment in ERP and other business solutions have been so overwhelming that the companies set aside this problem for another day—except that day never arrives.

Meanwhile, the ERP technologies have been evolving at their own frenetic pace, which adds to the challenge of figuring out what to leverage and what to ignore. SAP appears to purchase a company every six months, including such well-known brands as Ariba, SuccessFactors, Concur, Fieldglass, Business Objects, Sybase, and Hybris. It's hard work keeping all these disparate technologies current on a half-dozen separate and incompatible database solutions.

In a low-margin world, the functions are under intense cost pressure. The relative costs of complex systems rise, along with the costs of the functions tied to these systems. Cost improvements are hard to achieve because ERP systems were designed for a world that no longer exists. They predate many of the most important developments, which occurred in the past ten years.

ERP systems were originally designed for a desktop world with a big screen and a mouse, not for a world of smart devices and tablets. The system owner used to provision all the necessary computer infrastructure (machine rooms, networks, disk farms, or backups) internally, since cloud computing hadn't been invented. Original ERP systems predate features like Internet-driven security, browser interfaces, screen rendering, and modern user-experience design. ERP assumed that people were the least expensive resource and its system was optimized to ration disk use, memory, and processing power—the relative expense is now reversed. Even a license-per-user billing model doesn't fit a world where things may be the user, not a person.

An oil and gas company today faces enormous pressure to achieve a step change in cost and productivity but typically has hundreds of separate and incompatible systems, a big ERP investment with lots of modifications and workarounds that is costly to maintain, difficult to change, and not made for today's digital world. The ERP providers are themselves anxious to evolve their products to reduce the cost of ownership and take advantage of the incredible potential of digital.

The support functions have little option but to go along with the upgrades for the enterprise systems. They do, however, have some degrees of freedom to implement changes that could move the cost and productivity metrics more dramatically, but they must strive for the maximum possible value from their investment.

The Future State

As I see them, the support functions play an important role in ensuring high trust and confidence levels in the organizational context. HR promotes trust principally inside the company between employees and employer: that pay is fair, vacations are taken, and performance is captured. The supply-chain organization promotes trust between the company as buyer and vendors as suppliers of goods and services: that prices are correct, that goods meet specifications, and that procurements are legitimate. The finance function promotes trust between the company as borrower of funds and the capital markets as lenders and shareholders: that the costs and expenses are properly recorded and shared according to the rules.

Digital technologies play a key role in driving up trust but at lower cost and with fewer manual interventions than currently. Greater trust levels between parties means the elimination of many of the kinds of artifacts on which business has been based for as long as it has been keeping records. To illustrate this point, consider how a standard supply-chain process might be very different in a digital world (and, to be clear, all of these digital technologies exist today).

- A self-monitoring Internet-enabled tank recognizes, because of its own onboard sensors and analytics, that the tank is approaching full and requires a routine pickup and drop-off at a tank farm.
- The tank creates a smart contract on blockchain to record the event and alerts the company's ERP system. The ERP system selects an appropriate supplier of services, based on the supplier's performance record and proximity to the tank (the truck is conveniently visible online).
- The supplier's ERP system logs the request and accepts the work, updating the smart contract with the supplier pickup details (expected timing and costs) and sending instructions to the driver.
- The truck arrives to unload the tank, and the unloaded volume is captured by a smart sensor on the tank and matched to the volume loaded onto the truck, also via smart sensor on the truck's intake

pipe. Volumes match, and the smart contract is updated with volume, date, time, and carrier.

- At the tank farm, the process repeats. Once the truck tank empties and volumes match, the smart contract is updated, and the supplier is automatically paid. Then both ERP systems are updated to close the order, reflect the supplier performance, and record the financials.

In this example, no purchase orders were created. No pickup slips. No field tickets. No extra trips to the tank to check volumes. No invoice. No check. No phone calls. A considerable amount of cost melts out of the system.

This example repeats itself over and over, anywhere a measurement of fact could be captured throughout the supply chain by a sensor-enabled thing or by people with sensor-enabled gear. Machines will progressively take over the more mundane work and supply the critical data needed for fulfillment of obligations. Bots proliferate to execute wherever business rules can be codified as a set of instructions. Companies will have fewer but more important systems, whose interconnectivity with suppliers depends on cloud computing. The support functions should have fewer clerically oriented people than today, executing with much higher productivity and focusing the majority of their time on exceptions, not on the routine.

Underpinning this future world are the technologies of trust, including the next-generation ERP solutions, sensor technologies, and blockchain. A high-trust automated future is well within reach but carries with it important caveats. First, the world of sensors will take time to unfold and roll out, but make no mistake, it is coming quickly. Some suppliers are moving aggressively to embrace the Internet of Things. Second, next-generation ERP solutions can implement in multiple modes: by preserving the manual, artifact-driven, and costly administrative structures; by implementing computer-driven high-trust highly automated processes with key business partners; or by hybrid designs. Smart businesses would aim for the most efficient automated solution that involves sensors and blockchain, while allowing some of the more cautious business partners to adapt more slowly.

Finally, the data-rich world of finance, supply chain, and HR promises to be a lot richer but with greater accuracy and timeliness. The month-end frenzy will reduce in intensity, and the business as a whole will operate with the kind of precision associated with manufacturing.

KEY MESSAGES

This survey of the oil and gas value chain demonstrates the impacts of digital innovation and the pervasiveness of such technologies. Key takeaways about the impacts of digital on the value chain include:

1. All elements of the oil and gas value chain have opportunities to exploit digital innovation, and all participants in the value chain are equal in their ability to pursue digital solutions.
2. Digital innovation will expand supply dramatically, while helping to reduce cost and improve productivity throughout.
3. The changes in the transportation sector will trigger demand destruction, at first imperceptibly and then suddenly as the automotive manufacturing supply chain shifts over to new drivetrain technology.
4. The impacts will be biggest in gas retail, which is the most dynamic due to low barriers to entry and broader retail industry changes.
5. Digital enables asset-light business models to develop almost everywhere. Nontraditional competitors will be the biggest threat to operators.
6. A core of digital innovations consistently emerges across the value chain, including cloud computing, artificial intelligence, and the Internet of Things.

ORGANIZING A DIGITAL PROGRAM: Ideas for Managers

SO, YOU'RE CONSIDERING a digitization program in your oil and gas company, and you need to organize the effort. This task will certainly be on the minds of those who are either already working hard at the effort or thinking seriously about it. Perhaps surprisingly, how you organize for digital does not fundamentally depend on the part of the value chain that is your playground. You will face many of the same barriers, hurdles, and challenges as your colleagues across the sector.

Digital is confusing, mysterious, and tantalizing: it shape-shifts frequently, with new ideas and solutions. How about brainwave sensors in hard hats to detect sleeplessness? Done. What about crowdsourced geology interpretation? Happening already. Could we use low-flying drones to measure rogue methane emissions from gas wells? Being piloted. Bitcoin technology to automate the trade in LNG? Coming next week. Could we 3-D-printed replacement parts for field equipment? Trials are underway.

How do you go about organizing a digital transformation program when the field is so variable? Even agreeing on a definition of digital is a challenge. And there are a few significant ways that digital is very different now from the technologies of the past.

One of this book's interviewees suggested how valuable it would be to have a how-to guide for implementing digital. While not quite a recipe book, this chapter focuses on the essential ingredients for a successful digital effort:

- **Governance:** Where will decisions about digital investments be made?
- **Piloting:** Where will you start your digital journey?
- **Data:** How can you convert data into a strategic weapon as well as digital feedstock?
- **Cyber:** Are you prepared for the inevitable cyber incidents?
- **Talent:** How will you prepare your workforce to embrace a digital future?
- **Agility:** How do work practices need to change to align to digital?
- **Change management:** How will you persuade your organization to follow your vision?

GETTING GOVERNANCE RIGHT

Determining which part of the organization should drive digital innovation is dependent on its size, culture, organizational construct (such as centralized versus federated), and maturity. Ownership of digital innovation might be a fit with corporate IT, operating business units, a combination of the two, or a specially developed cross-functional team.

Driving Digital Innovation via Corporate IT

The launching point for many digital transformation efforts is to view it as a variant of information technology. Internal IT shops likely already have support processes for those ubiquitous digital devices, the smartphones and tablets (so that we can all bring our favorite personal device to work). Company-sponsored apps might run on these devices; the users may be able to access corporate IT services like email or personnel data in the same manner as on desktop devices, perhaps through a browser; and, in some cases, users may be able to access corporate-sponsored cloud-computing solutions.

IT departments in big oil have many of the key skills to lead digital transformation. IT likely operates a help desk for smartphone incidents, manages cyber issues like passwords and authentication, and mandates or limits certain solutions for anyone inside the corporate firewall. For example, my former company prohibits collaboration tools that lack built-in encryption because of the sensitivity of company data. IT is typically highly skilled in building and maintaining data-exchange integrations that connect the various business systems. IT will also have the infrastructure in place to support the wave of additional sensors, cameras, and monitors that are likely to be deployed in the future, which will need cyber protection common to today's smartphones, but not so common in legacy sensors. These capabilities include virus detection and remediation, patch rollout, version control, and change management of software across thousands of devices.

All this suggests IT is a good place to house the digital transformation team. But do not assume that because IT looks after the corporate compute infrastructure, it should also take the lead role in driving digital transformation. Corporate IT may have a huge workload already (many oil and gas companies have hundreds, if not thousands, of separate IT solutions that demand care and feeding). The work processes that many IT departments have in place to manage a very large and complex portfolio of technologies lack the nimbleness that digital transformation requires.

CORPORATE IT WILL often impose its own list of requirements for digital that can serve as barriers to digital innovation. Imagine this dialogue between the corporate IT buyer and the digital company (DC):

IT: Can you integrate with SAP and all of the thousand other systems we have here?
DC: We include an API set that connects to any cloud-based software.
IT: We don't trust cloud computing. Can we house your solution in our data center?
DC: Possibly, with extensive modifications and no ability to upgrade.
IT: Is your solution fully compatible with our standards that originate in the 1990s?
DC: Why is backwards compliance to the age of mainframes relevant?
IT: We don't have wireless networks in the field. Is that a problem?
DC: Why do you have people, assets, and equipment offline?

Driving Digital Innovation via Operations Technology

The big money in oil and gas is in operations. Most, if not all, of the physical assets and facilities belong to operations, and the vast majority of those assets predate the rise of digital innovation. Operations has invested mightily in continuous improvement initiatives, asset-management solutions, and supply-chain enhancements to drive better returns. Industry observers agree, however, that the biggest benefits to oil and gas from digitization likely reside in operations. In its 2017 report on the impact of digital on the energy industry, IEA estimated that digital innovation would impact operational performance in critical ways:

- a 7 to 15 percent reduction in energy use,
- a 30 percent improvement in asset availability, and
- a 90 percent reduction in unplanned asset failure.

Operations understands just how difficult it is to introduce change to plant and infrastructure. These installations usually run 24/7 with technologies that date back to the sixties in some cases, process unforgiving and dangerous materials, have very small and overloaded outage windows through which to introduce mechanical changes, and have multiple worker shifts to train up. For the most part, operations is measured on things like incident rate—zero is the usual target—reliability, availability, equipment utilization, on-spec output, optimal energy usage, and costs of operations. These factors place a premium on stability, carefully managed change, and as few interruptions to the business as possible.

Most oil and gas concerns will have an operations technology (OT) support team, whose focus is on such things as the field sensors and SCADA systems that monitor plant performance. A digital transformation program could be part of this team's mandate, but, like corporate IT, it's not a perfect fit. Operations technology is generally very stable, hasn't historically faced the regular upgrade cycles common to commercial IT, doesn't have the exposure to cyber issues as plants are often "air gapped" or fully disconnected from the Internet, and likely doesn't have as much exposure to cloud computing. The driver for OT work is often the mechanical attributes of a plant—unless the plant changes, there's not much need to change the operational monitoring capabilities—and, as a result, OT support teams lack the processes and procedures for introducing digitally driven change.

Driving Digital Innovation Separately

A third approach is to create a stand-alone digital transformation team that is neither in corporate IT, nor in operations, and task it with driving digital forward. This approach has some merit. Such a team would be free of the biases and challenges inherent in IT and OT. It could take a more clean-sheet approach unbound by the usual budget limits, standards compliance, and legacy business alignment challenges. Organizations that are particularly slow to change, or have such strong culture that change initiatives frequently fail, might consider this approach.

MANY DIGITAL INNOVATIONS will run into deeply entrenched business practices that unintentionally block creativity. Imagine a digital company (DC) having this dialogue with a big purchasing department (BPD):

BPD: Here are all our requirements; can you meet them all, immediately, and out of the box?

DC: Perhaps after the proof of concept, I might be able to meet some of them.

BPD: Do you have a list of satisfied customers?

DC: Our early adopters have been satisfied with the experience of working with us, yes.

BPD: Do you have a "product"?

DC: Not really, it's more of a platform, and it's minimally viable and it will likely pivot to something completely different once we start to work together.

BPD: What is your standard commercial arrangement?

DC: We're not sure how to charge for it yet. It's a two-market design with a blockchain overlay and an advertising-based cash flow.

BPD: What is your annual maintenance charge?

DC: I'm not sure what that means. The software will likely be free.

On the other hand, such teams could create delightfully elegant but thoroughly implausible solutions that cannot be easily adopted by either operations or corporate IT. Much like an architect who has only designed but has never built a house before, the team might not truly understand how the business's economics work or may end up focusing on the wrong problems. And a separation of the digital transformation team from the existing technology teams could create its own dynamic challenges, from resentment to envy or to feelings of inadequacy and of being punished for past technology issues and choices.

Driving Digital Innovation Jointly

Digital is so different from both traditional corporate IT and operational technology that a transformation team should include business insight and deep business understanding, a solid grasp of the nuances of daily operations, knowledge of existing business systems and device usage, and specialist know-how in specific technologies and capabilities.

Future high-impact solutions (like using drones to inspect and monitor gas infrastructure) rely on business, IT, and OT working very closely together. Drones are great examples of OT technology. They consume and generate real-time data about operating conditions, but they rely on commercial data systems for work-order history, parts inventory, and corporate reporting. As another example, the heavy hauler trials in the oil sands are deeply embedded in operations but need crew scheduling.

In my view, the most appropriate design incorporates these four sets of capabilities: business users, IT, OT, and digital technologies. All have specialized insights to contribute to how digital could impact the business. Their strengths are in fact complementary.

IT IS BETTER at cyber, but OT is better at reliability; business users are best at business insight and the best way to introduce change, and digital technologists know the strengths and limits of digital.

The specialist digital skills could be imported for the job, in key areas like user interface, user-centric design, scale effects, digital economics, and digital innovation, as well as in specific candidate areas for investment (e.g., big data, blockchain, 3-D printing, augmented reality, artificial intelligence, robotics, and new business models).

The ideal combined team would be composed of up-and-coming leaders from the organization and will ideally report to an executive tasked with driving digital innovation across the organization (that is, the executive might not be either the CIO or the COO).

THE FIRST STEP IS THE HARDEST

Once the governance structure has been determined—at the corporate level, business unit, or elsewhere—the organization needs to choose where to start experimenting with digital. If they haven't already, organizations will engage in a discussion about the opportunity and the need to react to digital innovation. However, I find this is often where organizations struggle. Digital is advancing in many directions, impacting much of the industry at the same time, and that complicates and muddies the discussion. Where should the organization focus its efforts and when?

My suggestion is just get started somewhere. Once you do, you will learn where your company places value. Not all of your efforts will be successful. If they are, you are doing something wrong, either by not thinking big enough about the potential changes that are coming or by focusing solely on incremental business improvements.

Because each organization is different and digital is expanding exponentially, there is no one-size-fits-all solution suitable for all players across all industry segments. However, there are some framing questions helpful to these discussions: Do you focus on operating assets or overhead; commercial or operational systems; short term or long-term orientation?

Operating Assets versus Overhead

As outlined earlier, there is a substantial prize to applying digital to existing, or brownfield, assets. Recall that brownfield oil and gas infrastructure is the equipment that is already installed and operating.

“I have zero interest in applying digital technologies. I have lots of interest in saving money.”

A CIO OF A LARGE MULTINATIONAL UPSTREAM PRODUCER

Greenfield, or brand-new, is still on the drawing board. It is much less costly and simpler to make digital changes to a greenfield design than to an operating (brownfield) plant.

A brownfield plant is usually energized with high voltage, or it may be heated, pressurized, or have rotating equipment such as pumps, which means the plant has to be switched off, cooled down, drained, and de-vaporized before work can be done on it—including making digital changes. Shutdowns of this nature are infrequent. A comprehensive shutdown of a complex plant could take a year to plan and the shutdowns may even be years apart. It may well be impossible, or at the very least impractical, to test a digital adaptation in a live and running plant.

Oil and gas is an inherently brownfield business. Almost all production is coming from older assets that predate the dramatic rise in digital capability and were not designed for the digital age. According to BP's annual statistical review, the global oil and gas industry produced 91.6 million barrels of oil (mbbls) and 274 billion cubic feet (bcf) of natural gas every day in 2015. The vast bulk of that production comes from facilities designed and built before the oil price crash that started in June 2014.

Thanks to the collapse in oil prices, capital spending was severely constrained. Various industry analysts calculate the capital shortfall as being in the hundreds of billions of dollars between 2015 and 2018. There are very few, if any, big, new, and modern assets that have been designed and built from the ground up with digital in mind.

Two areas that will see quicker uptake of digital are light tight oil (LTO) and steam-assisted gravity drainage (SAGD). LTO, referring to the low viscosity or light petroleum found in low permeability shales and sandstones, can embrace digital most rapidly, since these plays are relatively low capital, compared to, say, a new offshore platform. LTO wells deplete quickly enough and require a constant flow of new capital to keep them productive. Applying digital innovation could work exceptionally well because each new capital iteration presents a fresh opportunity for digital experimentation. The same holds true for coal seam gas wells, shallow gas fields, coal bed methane, and SAGD wells.

Oil and gas companies need to understand their unique mix of capital and operating costs to guide digital spend. For example, in the upstream, those heavily weighted to brownfield steady state (oil sands mines, large conventional fields, or offshore production assets) will likely orient their plans to preserve value by making incremental improvements to their operating cost structures. Those upstream players heavily weighted to greenfield capital spending (SAGD and shale players) will likely orient their plans to achieve more dramatic reinvention of their capital and operating costs. Downstream-only players have more challenging choices because digital's impact will be greater and the possibilities of value loss more pressing.

How much priority should be given to reducing general and administration costs (in areas like HR, finance, IT, and supply chain) over efforts to transform operations? The general and administrative (G&A) cost structures for the oil and gas industry, with perhaps the exception of the very largest players with scale economies, have been more or less static for a decade or more. In my view, the prevailing industry business model (functional specialization and asset-driven) appeared during the previous big wave of mergers (Exxon and Mobil, BP and Amoco, and Chevron and Texaco) in the 1990s. The model has changed only slightly since then, with the adoption of global service centers to house these big G&A costs.

Fortunately, digital technologies are more mature in the G&A areas, which are common to all industries, and so payback from digital change is probably faster and lower risk than in operations. I see some big juicy targets in G&A after a decade of relentless regulatory change, proliferation of systems, and a recent emphasis on growth in the era of $100 oil. At least one large global oil and gas company has more than 10 percent of its head count in finance.

Of course, focusing on G&A costs can only go so far. Eventually, oil and gas companies will have to orient their efforts to reducing operating costs and growing production. In the meantime, G&A costs should get some attention.

Commercial versus Operational Technology

What should be prioritized for digital enhancements: IT or OT? Much like how production assets are brownfield, the same is true for the thousands of legacy information technology and operating technology systems installed across the business. Many (if not all) of these systems predate digital's current opportunities (low cost computing, unlimited storage, mobility of people and assets, autonomy and robotics, and ubiquitous networks). These two domains (IT and OT) have operated as separate computing environments in oil and gas since computers were invented, but today there is pressure to upgrade both domains.

In the main, OT systems are the more static of the two. OT systems directly control and monitor pressurized and heated units, run 24/7, have very high uptime requirements, stringent safety features, high operating reliability, and fail-safe needs. These plant systems cannot be changed once they're switched on without bringing the plant down, and bringing the plant down has a direct impact on production volumes and revenues. Listen to any upstream investor call, and the CEO will invariably start off with a summary of the quarter's production.

IT systems are the more fluid. They change more frequently and are constantly being patched with the latest feature set or security profile. But the CEO rarely mentions IT issues in an investor call unless they have been victims of a cyber-attack.

The problem is that digital solutions for oil and gas increasingly look like they will require IT and OT systems to become more integrated. Take aerial drone technology, available today from a few suppliers. A proven use case is drones that fly over operating oil and gas wells, taking measurements of vegetation, moisture, damage, emissions, and operating state. The pilot who prepares the flight plan would be guided by IT systems containing asset data, maintenance history, and well configuration. Actual flight operation is a classic OT system, which might feed data directly to operations about real-time well status, as well as to IT systems to auto-generate work instructions and shift planning for well services.

It is becoming increasingly clear to chief operating officers and chief information officers that digital advancement has numbered the days of this arbitrary duality. Oil and gas companies will need to be selective in allocating investment dollars and upgrades in both IT and OT areas, and most will likely need to figure out how to bring them closer together in a converged business model.

Short-Term versus Long-Term

With so much uncertainty around oil and gas markets, it's tricky to decide how to balance short-term cash benefits with longer-term strategic goals. On the one hand, if prices rise, I would rather have capital at the ready for growth. On the other hand, if prices stay flat or decline, I would rather work on getting operating costs permanently lowered.

There will be voices internally advocating a do-nothing stance to see either commodity prices turn around or clear winners and losers to emerge in the technology race. The default position for many oil and gas companies is to wait, let someone else take the lead, and aim to be a fast follower.

Doing nothing, however, is not a sound choice at this time. Prices look very much like they will stay at moderate levels for the foreseeable future. More worrisome to me is that the bulk of recent industry cost savings has come from the supply chain. Oil and gas companies have squeezed their suppliers hard, but suppliers tell me that they have simply cut their prices while removing a bit of capacity from the market. Should oil prices rise, suppliers' prices will rise in tandem. The actual work carried out in the industry has not become more efficient following 2014's downturn in commodity prices.

Oil and gas companies should be investing now in new ways of working, both internally and with their suppliers, to keep costs out in a more sustainable manner. Digital solutions could play a big role in achieving this through process automation, improved data quality, better collaboration with the supply chain, and deeper analytics.

Until market and technology directions are clearer, getting the investment portfolio balanced properly requires a structured investment plan that has immediate payback items, such as process automation; some foundational investments, such as data quality improvements; and some deferrals, such as automated vehicles.

TACKLING THE DATA MONSTER

The next priority for a digital transformation team is to ensure that plentiful high-quality data (one of the core building blocks of digital) is available. Data feeds analytics, and without good data, analytics are likely to be incorrect, inaccurate, or untimely. The oil and gas industry is similar to a number of others that have discovered that data assets are actually critical, thanks to the rise of the data-intense digital companies like Google and Amazon. The way the industry has treated its data to date is ill-suited to a digital world that features rapid growth in data quantities, different kinds of data, and novel analytics.

Data Accountability

Accountability for data in oil and gas is diffuse and shared, which creates challenges in this era. The most senior executive charged with managing commercial data is usually the chief information officer, who typically reports to the CFO or to the senior vice president of corporate services. Their role is generally confined to the IT side of the business (ERP, trading, email services, and phone). They run the data center and networks, adapt to technology advances, deliver new business solutions, and provide cyber security.

The chief operating officer will be accountable for the operations technology (OT) that runs the facilities that extract, process, and transport oil and gas along the value chain. Some oil and gas businesses are more asset-driven, with asset managers who are responsible for OT for their assets. The VP of exploration will be accountable for all the geology and subsurface technical tools for interpreting such data.

Accounting rules play a role in determining how oil and gas companies treat their data, and value drives investment in and behavior towards that data. For example, is data "free," as in "no cost"? Or is data valueless? An interesting example is seismic data, which is usually recorded as the cost to recapture the data and not necessarily the value that could be released through its effective analysis.

In general, oil and gas entities do not record their data assets on their balance sheets with a value. Data is more associated with being a cost or an expense, recorded on the profit and loss statement as an IT cost; it is managed on the basis that the cost be kept to a minimum. Expense items struggle to attract capital investment and talented people, and during the commodity down-cycle, expenses come under intense cost pressure.

This much diversity in accountability hinders the consistent management of data, introduces security weaknesses, and makes analysis of the business more difficult than it needs to be. It also blocks clever digital initiatives that derive value from combining multiple technologies across multiple participants in the value chain. Consider the example of a blockchain solution that captures emissions for trading purposes. To work properly, it needs access to both operating data (asset, location, operational output, and emissions measures) and commercial data (carbon-tax rate, emission owner, and emission balance).

While capital markets and accounting rules may not favor highly formalized valuations for data assets, managers can always create their own internal accounting metrics to promote the right kinds of behaviors.

Data Origination

Much of the useful data for oil and gas companies actually originates with suppliers (technologies, assets, services, and equipment). It's their sensor-enabled equipment that creates data. However, suppliers often have structural disincentives to provide open, standardized, easily accessed data for their customers out of fear that their sensor and equipment offerings will be commoditized. Indeed, many suppliers aim to sell fully integrated offerings, consisting of tightly coupled product families, which lock in their customers.

These "closed" systems may not have the software interfaces for easy integration with other systems. For example, a down-hole pump equipped with sensors may throw off lots of useful data, but that data is usually in a proprietary format and the pump controllers lack the software smarts (either intentionally or unintentionally) to allow access to it. Closed systems prevent infrastructure owners from adopting best-in-class components and leading-edge technology.

Next-generation buyers of oil and gas technology are challenging the entrenched suppliers to the industry to open up data models and enable better information access. I believe future procurement specifications for control systems will demand that they be standards-based, open, secure, and interoperable. Digital initiatives will increasingly need to extend beyond the fence to include the digital efforts and solutions from key suppliers.

"People somehow think that these digital tools will magically go into the crap they have for data and pull out results. The problem is the data is fragmented, poorly structured, and in thoroughly inconsistent formats. People build their companies around work processes, not a data model."

A CIO OF A LARGE MULTINATIONAL OIL COMPANY

Reliability, Accuracy, Availability

Oil and gas has many legacy practices that make data hard to work with. First, the data is stored in separate departmental or functional silos, in various incompatible formats, with technologies and solutions selected with narrow terms of reference. Frequently, the data is captured as an afterthought rather than an integral part of the business process, so there is little concern for anomalies or inaccuracies. Errors in the data make it less trustworthy. Over time, measurement devices can drift out of alignment with real conditions and require recalibration, a costly exercise.

Engineers then spend considerable time pulling these large datasets together, trying to overcome its errors to improve its reliability, and sharpening its accuracy to reflect real operating conditions.

Throw in a bit of organizational politics that slow down the adoption of enterprise solutions, add the can-do and independent engineering culture, and, not surprisingly, oil and gas datasets duplicate and multiply with abandon. Determining which dataset to use is like confronting a Jenga tower of data—you can pull out a dataset to play with, but you never know if the whole thing might come crashing down.

Variety and Volume Acceleration

Oil and gas data shows the same two growth vectors as in other industries—tremendous variety and enormous volume. These two vectors, variety and volume, have always shown robust annual growth but are now accelerating thanks to digital advancement. Next generation technologies produce more unstructured data (more variety), such as autonomous kit with its cameras, audio recorders, and LIDAR. As sensors fall in price, they will appear on more kit (more sources) and generate yet more data more frequently (more volume).

I draw an important distinction between "large data" and "big data." A large but well-structured dataset in a tabular form is a challenge to process but not because the analytics pose a challenge. It's more about scale and pure processing power. "Big data" is unstructured data, like text messages or pictures or music tracks: there's no inherently obvious way to relate text messages to pictures to identify patterns and trends. Big data has both scale and structure challenges that inhibit its analysis.

Oil and gas is already proficient at collecting and consuming large data. There are even pockets of capability in big data (think seismic and map processing). But most oil and gas executives will readily admit that their companies are not very good at exploiting big data, or analyzing it for insights and business improvement. One of North America's largest oil producers estimates that they use at best 0.5 percent of the data that they have amassed over their years of operations.

Data Strategies and Tactics

Faced with a large and rapidly growing mountain of data of uncertain origin, accuracy and reliability of which are questionable, but with promising digital tools coming quickly, where does one start?

If data challenges were just technical in nature, companies could simply purchase a fix or develop a hack and be done. But, as I've set out, the challenges are much more complicated, and the issues are devilishly interrelated. Companies need a more strategic approach. Data issues must be tackled holistically so that changes in one aspect (such as an agreement to adopt data standards) are not in conflict with another (such as procurement policies based on least cost).

A good strategy for dealing with data would include a survey of the key issues surrounding data, a view as to what "good" data management looks like, and the leaders in dealing with data (and the leaders might be outside the industry). The strategy would:

- set out key goals and objectives for data, such as high reliability, low duplication, and clear accountability;
- define a set of investments, such as technical, standards, and procurement, that are required to achieve the goals;
- deliver an organization and the resources to own the data initiatives; and
- create a set of metrics to monitor progress.

With so much data available to work with, a good step is to cut the mountain of data down to a workable size. This means quickly finding the data that matters.

START WITH THE development of a value-driver tree, which is a representation of value (revenues, costs, and assets) and what drives that value (increase in price, growth in volume, or greater throughput). This economic view of the business helps shed light on precisely how value is created and what expectations shareholders have built into their valuation of the business. Best-in-class value trees include the levers of value and the tradeoffs between such inputs as machines, labor, energy, carbon, and water.

Once the drivers of value are clear, a value assessment can be placed on the data required to manage the driver. Managers can then properly allocate capital and talent based on the value of the data. Existing datasets that support the drivers of value will surface through use.

In my experience, oil and gas companies are not short of analytic tools. In fact, odds are there are likely too many tools from which to choose. Diversity of tools is not always a good thing—clever models built in one tool may not be leveraged in another tool. Consider putting the tools into a tool library and let the power of crowd-usage identify those of greater value.

Another critical time-saver is to make data assets more accessible. Instead of hiding them away on server drives, departmental systems, or inside ERP systems, datasets could be stored in one accessible place, accompanied by their descriptive data (or metadata). Metadata is key as it enables searching, filtering, building relationships between datasets, labeling charts, and providing references. In time, new data could be made to conform to agreed data standards (assuming that your organization can agree on standards to be followed).

Industry leaders are trialing the creation of these data lakes, which are single online and searchable repositories of all data assets in their original formats. I think of a data lake as a database of databases and datasets, ranging from the original, raw data formats to more formalized spreadsheets, and more. The best data lakes incorporate a navigation map to the data they contain, using metadata to describe all the columns and what they mean; access to the host of analytic and visualization tools available; a way to search for content; useful templates that others have shared; and "like" buttons so that the best data and tools surface for others to use.

The analytic possibilities presented by linking different datasets from different parts of the business are endless. For example, one company took all the GPS data from its trucks and its suppliers' trucks (generated while driving to and from gas wells) and overlaid that on a revenue, cost, and production view of the business. They discovered that the high cost of well maintenance was in part caused by all the trips that the engineers needed to make to the wells. The reason for the trips was driven by the poor quality of information about the wells back in the home office. The engineers could not rely on that information to make effective decisions and were compelled to personally visit the wells each time.

In another application, an operator merged health and safety incidents onto its human capital data and learned that the biggest predictor of incidents was the average age of crews and shift length—the older the crew, the more likely they would experience an incident as they neared the end of a long shift. In a second comparison, an operator discovered that its incident rate peaked following longer seasonal holidays where certain workforce segments returned home to family. Long days of celebration and revelry, followed by short nights with limited sleep, yielded an exhausted returning workforce that suffered overly high incident rates.

Finally, consider the possibility that the data about oil and gas reserves and deposits is actually the real value driver. In the same way that capital markets reward data-centric and asset-light business models, several companies are eyeing the potential to become the data depository for the whole of industry and to apply AI tools to that data. Some of the largest cloud-computing customers are oil and gas companies, which recognize that the value of their data may be in how different data are integrated within the company and with other players in the value chain.

MANAGING THE CYBER RISKS

Now that you're getting a grip on the mountain of data assets, it's time to check into your cyber-security preparedness, which includes deterring cyber-attacks and keeping data protected while staying current with regulatory changes and having a ready-to-go response plan when the inevitable cyber-attack happens. Does your organization know who to call when an attack is detected? Are there procedures in place to limit the damage? And when do you alert your stakeholders?

Data and Cyber Security

Hardly a week goes by without some ominous news story about yet another cyber-attack. Yahoo! had some 3 billion names, email addresses, and passwords stolen. Equifax had financial records from 150 million customers compromised. Even Uber, a modern digital company, has been attacked, with records of some 57 million customers lifted. Computer viruses like WannaCry (a piece of code that threatens to erase critical data unless a Bitcoin ransom is paid to an untraceable account) have hit many companies around the world. Even holders of digital currencies like Bitcoin and Ethereum have been successfully hacked, with digital wallets emptied.

Most of these stories involve the theft of personal data (such as individual email addresses and passwords). Hackers like to target large companies because it takes the same amount of work to steal one email address as it does 150 million. Hackers like personal data because it's easy to sell and exploit. For example, in the case of the Equifax hack, enough personal data was lifted from the company's databases that nefarious types could set up fake bank accounts, arrange false loans, and authorize thefts, all under the names of real people.

Few companies in the oil and gas industry would appeal to hackers interested in stealing personal data. In many countries, there are just a handful of petroleum retailers big enough to have millions of customers. In some cases, that relationship would be through a third-party loyalty program (the kind where you insert a card to identify yourself to the pump, which then tracks your purchase).

The Boards at these big players have challenged the VP of downstream and the CIO to get ahead of the typical hacker armed with usual range of attack vectors and objectives, including accessing customer data, extracting value through ransomware attacks, achieving a distributed denial of service (DDOS) attack, or redirecting compute power for personal use (such as to house pornography, mine for Bitcoin, or attack other computers).

Infrastructure Vulnerabilities

The vastly greater threat from cyber activity in oil and gas is in the production infrastructure.

Almost every oil and gas well in production today has sensors and actuators connected to a SCADA system. These sensors collect data from the well in real time—that is, every millisecond or so, sensors collect a little data (pressure, temperature, or speed) and send it to the control system, which decides what to do, such as open or close a valve, and that data is also displayed on a screen, maybe as a graph or as an indicator, in a control room somewhere. Since the wells are spread out across vast geography, so are the sensors and systems. A new term, edge computing, refers to these kinds of systems where a remote device has some computing capability.

It matters how edge systems are connected to each other. Oil will flow from a well controlled by one SCADA system into a gathering system for many wells that may be controlled by separate SCADA system. The gathering systems may connect to batteries or tank farms or pipeline systems, each of which has its own supervisory system. It's not at all unusual for a large oil company to have hundreds of separate SCADA systems looking after thousands of wells, which have been bought and sold over the years.

If a hacker broke into one of these sensors, they could send bad data to the SCADA system and trick it into opening and closing valves, raising temperatures, boosting pressures, or cutting power. A cyber-attack could even cascade through the systems if they were systematically compromised. Imagine the risks: potential damage to the environment, possible harm to employees and contractors, impacts on residents nearby, the possibility of damage to assets, and impacts on shareholder value.

Why not simply upgrade to one massive integrated SCADA system for all equipment and sensors? It costs a lot of money to retrofit production assets to a new system with little operational benefit, so most operators leave whatever system came with the asset in place. Some wells produce so little oil that it would make no economic sense to upgrade at all. Other critical assets, like tank farms, run 24/7 and are part of a continuous flowing business, so taking them offline to replace systems that keep the assets running is virtually impossible.

This production infrastructure at the edge is a very attractive target for hackers for several reasons:

- These sensors and SCADA systems are old—they may date back thirty years or more, well before the rise of widespread cyber activity—and lack the tools to identify and repel attacks.
- The systems are defended by an "air gap" and were not designed to be patched like modern systems. There was no reason to include patching as a feature of the original design because they were not exposed to the outside world by cable or wireless and viruses hadn't been invented yet.
- The passwords to gain access to these sensors and SCADA systems may be hard-coded, which means they can't be changed. Get the password, which may be available in some online documentation, and *voilà*, you're in.

The obvious solution is to preserve the obscurity of these systems. Fortunately, many still do not connect to the Internet, and oil and gas companies and their suppliers don't publish which systems control which assets, so it would take a lot of work to identify meaningful targets (perhaps a job for robotic cyber software?).

But as time goes on, obscurity is becoming a questionable strategy. Modern sensors added to old SCADA systems may be directly connected to the Internet and create new vulnerabilities. New SCADA systems often connect to the Internet to enable new business models like single control rooms, direct supplier monitoring of key components like turbines and pumps, and access to all that data.

Ken Dick, a research fellow at the University of Nebraska, put a new fake SCADA system onto the Internet (as if it were a new oil asset) with some software to monitor how long it took the SCADA system to be discovered by robotic software on the web looking for such things. No surprise—it took merely minutes for a bot to find the system and start to attack it. The same effect has been noted for other devices added to the Internet, such as toasters and fridges.

With the rise of the Industrial Internet of Things (more sensors on more things, generating more data, and communicating that data to more computers), the attack surface is getting bigger. Gartner estimates that more than 11 billion sensors are on the Internet in 2018.

To fix this problem, we have to first elevate attention to the risk. At a corporate level, the risk matrix (where oil and gas companies set out what they see as the biggest risks and the probability that they might occur) needs to show cyber concerns much higher and to the right—more impact, more likely to happen. Addressing cyber concerns allows companies to preserve value (as with avoiding safety problems), whereas most other digital initiatives are about creating value (improving the efficiency of the business). Until the matrix is updated, cyber gets limited attention.

Next, we need to approach two fields of play—the brownfield assets and the greenfield assets.

For brownfield assets, the most worrying point of access are the edge devices—the sensors out in the field. There are many more of them, and they are more vulnerable. Industry needs to rethink the sensors' software to encrypt the data generated (which would thwart attempts to intercept the data and corrupt it) and enable the usual suite of capabilities for device management:

- Authentication—is this sensor authentic and not a fake? Is it recognized by the SCADA system? Has it been compromised in any way?
- Authorization—is this sensor permitted to exchange data or perform the task at hand?

Unfortunately, back when many of the legacy sensors were designed, cost constraints typically limited its amount of computer memory. There usually isn't room to add new software, particularly the industry-grade encryption software we use on our smartphones, tablets, and desktops. The sensors usually have limited processing power (fit for their task) and not much power to run the processor to do encryption work (an overhead task).

One interesting solution comes from AgilePQ, which has approached the problem with a digital lens. Rather than replacing the sensor outright, a costly answer, its software solution solves many of the issues of memory constraint, power usage, and processing limitations, while delivering robust encryption and reducing the potential for a bad actor to hack into the sensors. Cracking today's conventional encryption technology is akin to finding a particular grain of sand on the earth. Cracking AgilePQ's state of the art encryption is more like finding that same grain of sand in fifteen universes.

Industry should change its procurement standards so that greenfield sensors come with industry-grade encryption capability, support for patching and upgrades, and the usual array of capabilities for authentications and authorization. Market constraints should not be used as an excuse—suppliers are only too keen to include greater functionality in their solutions. Suppliers realize they need to address these vulnerabilities.

The Imperative for Change

Embarking on a digital journey exposes the company to more cyber issues. More digital devices mean more points of attack and of potential weakness. Concerns about cyber are now a frequent topic at the Board level. Regular media reports of cyber events have convinced most senior leaders that cyber-attacks have moved from possibility to probability, and that preparation for dealing with a cyber incident is now a business requirement. Businesses need to be able to identify and ideally prevent incidents, respond to incidents, and inform stakeholders and regulators of the impacts and remediation. How an organization protects itself from cyber events, how it responds to a cyber event, and how it preserves value are entirely within its control.

It is all too easy to slow walk, delay, or cancel outright a digital transformation because of cyber concerns. Unfortunately, the introduction of vulnerable digital technologies into the sector is already well underway—many suppliers of technologies now offer full Internet access for their devices as a standard feature.

Digital initiatives, regardless of size, scale, and focus, must incorporate some risk analysis of data-protection challenges, the possibilities posed by cyber events, and the regulations regarding data privacy, such as the European Union's General Data Protection Regulation (GDPR), and build in mechanisms to protect and preserve value.

THE TALENT MODEL

In our digital world, the talent model has evolved beyond where oil and gas companies are today. Not only will work be completed differently (with tablets instead of clipboards, drones instead of trucks, by exception instead of by routine) but the skills needed in the workforce are changing. Employees with a long history in the company, but with ten to fifteen years before retirement, will be more valuable if they are trained with needed digital skills, rather than simply packaged out when their role is made redundant.

The Future of Work

I have a clear sense that jobs are changing rapidly thanks to digital, but what will be the future of oil and gas work in this increasingly digital world? There have been a number of studies that suggest many jobs will be automated by digital, but nothing specific to the oil and gas industry.

In late 2017, I was involved with a sector study to understand where oil and gas would aim its digital automation dollars. Perhaps the best targets would be where the work is extra dangerous (and so elaborate safeguarding protections are needed). Perhaps it would be high-volume routine work (where automation could more easily replace human workers). Perhaps automation would be aimed at the extra costly work (by virtue of its location or by the scarcity of required skills).

Based on the research, there was almost no difference in the likely impacts of digital on these three categories of work, but there was considerable difference where digital innovation has made an impact already. High-volume routine work has received high attention and the deepest penetration of digital innovation. Examples include the application of robotic process automation (RPA) in finance and accounting, and the arrival of automated heavy haulers in Canada's oil sands mines. Other related heavy-asset industries, like Australia's rail industry, have developed fully autonomous trains. European truck-makers are piloting autonomous trucks that drive in platoons (where the lead truck controls the two trailing trucks). China has completed trials of fully autonomous bullet trains.

ROUTINE WORK

There are a number of reasons behind the selection of routine work as the digital target. First, the more hazardous work and the work that is most costly to do is typically dependent on investing more capital, but the industry is still capital constrained, making digital investment less economic. Second, some kinds of work that could be impacted by digital might be contingent on available infrastructure (such as telecommunications). Third, digitizing routine work tends to have the fastest path to value (i.e., the returns can be realized within a year). Time to value is a consideration in every industry, but is particularly important in oil and gas today.

I suspect there is low-value routine work in virtually all industry sectors, which in turn means that no sector can rest comfortably under the delusion that digital is some passing fad to be ignored. From the drones in Amazon's warehouses that replace forklift drivers and stock-checkers, to agricultural monitoring drones that replace helicopters and crop managers, to Tesla's self-driving trucks that deliver goods—if a job involves holding a steering wheel, there's a good bet it won't exist in a decade. It will take a few years for these technologies to reach tipping points, but once they do, industry adoption will be quick. Unlike those die-hards who cling hopelessly to flip phones, businesses show no such nostalgia. A fifty-five-year-old truck driver today will probably still be driving in ten years, but a forty-five-year-old truck driver will almost certainly be out of work before retirement and might need new job skills.

HIGH-VALUE OFFICE JOBS

Digital is also transforming high-value office jobs, essentially eliminating routine brain work, like reading and interpreting contracts, building standard spreadsheets, and writing narratives. Back-office teams of production accountants, cost accountants, paralegals, and land managers are being replaced by robots. I worry that some of these digital innovations will aim disproportionately at the women in the workforce in oil and gas, an industry that already has a significant gender bias.

Digital changes will affect the broader society. Public sector services like business licensing, taxpayer identity, and regulatory compliance will be reinvented through digital-enabled self-service business models, resulting in a reinvention of public administration.

The research showed that while it was less clear precisely when jobs in these sectors will be impacted, there was no doubt that the impacts would be dramatic and sudden. We concluded that transportation and logistics are likely first to feel change in a big way.

It is also dawning on the workforce that these digital changes will impact them. They read vague references to job displacement in the papers. They can see the impact of Google on search, Amazon on shopping, and Apple on phones. They note the bankruptcies of retailers unable to adapt fast enough. They can hear a fuse sputtering, but it isn't clear if the fuse is headed in their direction, how long the fuse is, and how big the bang will be. As a result, most are not preparing for the future in a thoughtful way. Even HR departments do not see the challenge or the impending job dislocation. I find most people go along with gradual changes in their environment but dislike abrupt change, and digital promises a lot of abrupt change. For example, Uber still triggers protests by the taxi industry in many cities.

Governments and education institutions are also struggling to discern what the future will be and to plan accordingly with learning programs and retraining assistance. STEM (science, technology, engineering, and math) skills, which exist in abundance in oil and gas, look like they are getting disintermediated by digital technologies! Oil and gas must anticipate that much of their workforce will be unprepared for the future, uncertain how to cope with the change, and unwilling to readily embrace digital innovation. I anticipate a huge step up in retraining and skills-acquisitions programs for workers.

The Human Hustle

If all work as we know it is going to be disrupted to one degree or another, then what is to become of the human worker? Put another way, what human attributes will be least impacted by digital's march into the work world? Certain attributes distinguish us from machines, but I should point out that digital innovators are likely working on emulating even these attributes in software and silicon.

- **Leadership.** The complexity of decision-making might be reduced by machines, but humans look like they will continue to own the leadership roles. That doesn't mean all leaders are safe—the shift supervisor of haul truck drivers disappears when the haul trucks no longer have human drivers, but reappears as a team leader for a squad of remote operators or as a robot controller. Media reports from China suggest the country, as the biggest buyer of industrial robots, is chronically short of robot operators. An assembly line can churn out an industrial robot every few hours, but training a human supervisor could take months.

- **Empathy.** Being able to connect emotionally under a broad range of settings (think of the act of selling, providing coaching, delivering bad news, negotiating land access, and addressing incidents) is a distinctly human skill. The future of work still looks like it will place a premium on relationship-building know-how, though AI technology may be able to help improve empathy in some settings (such as call centers and customer service).

- **Teamwork.** Forming teams with a wide variety of complex skills in short bursts to work collaboratively on tricky problems will remain with humans. I have trouble envisaging the creative process of brainstorming and invention being displaced by digital technologies anytime soon. I can absolutely see how some teams could be made more effective by delegating mundane tasks (like research) to machines. At the same time, traditional teamwork in factories is disappearing. Visit a shop floor—there are far fewer people about because the machines are dangerous. You might notice that the machines are often caged, and for our safety.

- **Communications.** Basic communications are now being put together by digital solutions, but the ability to tell compelling stories is inherently human, as it draws on the creative process. Tools like Quill (a natural language generator) can take raw data and compose a matching narrative, and these tools will only become more capable over time. Composing a message could be enabled by an app or cloud digital service, but the in-person communication of the message, such as developmental feedback, may well remain a human-only capability.

- **Customer insight.** Machines are not yet able to draw deep conclusions about human motivations and personal drivers. The predictive ability of algorithms is constantly improving and will get better as more data becomes available. But to truly understand why people react differently to the same stimuli, make radically different choices based on the same information, or behave differently under the same circumstances requires human intelligence. How do you predict heroism and radicalism?
- **Innovation and creativity.** Today in all fields, humans are the supreme innovator, able to imagine new solutions to problems, invent new devices, and create art. Machines and algorithms have shown remarkable ability to improve on human design but not to create breakthroughs. Humans are needed to design user experiences, advertising and promotion, and other creative roles.
- **Sensing and judgment.** People are uniquely qualified to sense when something doesn't quite feel right about a situation or to perceive inconsistencies that could be signs of fraudulent activity or bad actors. Facial recognition technology may be able to pick out a face in a crowd, but only humans can read subtle signs of someone behaving with deceit. Likewise, a computer could be programmed to evaluate a business case and produce a value, but the intangibles, such as the public relations impact from closing a factory, are hard to model and will remain with humans.

Our Learning Future

One of the key digital innovations to develop artificial intelligence involves machine learning. At a simple level, machines learn through repetition. Group a thousand different pictures of workers appropriately dressed in safety gear into a category called "safe workers," and you can teach a machine with an optical lens or scanner to recognize a worker not safely attired.

Digital technologies combine to create machines that are phenomenal learners. The autonomous trucks in the oil sands have an infinite capacity to learn, don't take breaks, and can be programmed to stay on task forever, or throughout their lives. They have what we might call high levels of self-motivation. These machines are connected to each other, so that lessons learned by one can be instantly shared to the fleet of other like machines (called fleet learning). We connect our digital machines back to their makers to keep them technically current. We protect them with cyber firewalls and ensconce them in data centers.

By contrast, humans spend 15 to 20 percent of our entire lives in schools trying to learn what other humans have already learned. For humans to maintain our place in the future of work, we must also adopt these same attributes. We must maintain our curiosity and be lifelong learners and create the support we need—sabbaticals, work tours, apprenticeships, job swaps—throughout the work world. We need to bolster our self-motivation to this task, because it is solely in our self-interest.

Tomorrow's Graduates

Only a handful of years ago, petroleum engineers were in such short supply that they could command some of the highest starting salaries straight out of college. Given what you know now, what would you suggest that a high school senior or university student consider when choosing courses to study or a career path? In a more digital world, for example, the kind of skills required in geology will change. Schools will need to incorporate more data science, more artificial intelligence, and more coding alongside studies of the Jurassic and the Devonian. These blended curricula will become the basis for success in the future.

I also expect to see oil and gas companies take on many more new hires outside of the engineering world, with job titles like data scientist, robot-management specialist, and user-experience developer.

AGILE: HOW DIGITAL GETS DONE

When I started my career with Imperial Oil in Toronto, I was taught that there was only one way to deliver computer technologies for users: the waterfall method. IT analysts interrogate unwilling and confused users to understand what they need by way of automation, using imprecise language and quirky documentation methods. These requirements are thoroughly hammered out before being handed to software developers to start coding, a process that could take months or even years. Eventually, the coders emerge and present their creation to the surprised user (a different person after such a long time) who says something along the lines of "That's not what I want," or "That's nice but the business has changed." This reaction kicks off fresh and protracted efforts to remediate the software without throwing it entirely away, leading to contracting woes or even lawsuits. Every few months, there is a media story about a $100 million software project that has not worked out, resulting in recriminations and litigation.

The waterfall process did work for technologies of the past. It made sense to deliver detailed specifications to the coders, so as to fix the scope of work, which would minimize demand on the scarce bits of the business. Fixed scope, in theory, would drive cost accuracy and schedule reliability, provided the requirements were accurately captured by the analysts. Software developers were highly paid and in short supply. Hardware was typically a mainframe computer with a dumb terminal, which had limited capacity and was expensive and slow. Networks and disk storage were equally costly and rationed carefully. Waterfall also worked reasonably well where business processes were highly regulated, users had limited flexibility in how they work, and business processes changed infrequently, if at all.

But the waterfall method creates its own hazards. The process is front-end loaded, and poorly defined requirements create lots of later risk. Users of technology don't often know what they want, and so the requirements can be wildly inaccurate. All the feedback comes at the end of the process, leaving few ways to correct the direction of the project if it's heading astray. The waterfall approach is not well suited to projects where the user lacks experience with automation.

It will not be lost on a heavy-asset intense industry like oil and gas that the design and build of processing assets, such as gas plants and wells, are done in a waterfall process. It's called stage gate, but it's the same approach. Requirements are hammered out in pre-FEED (front-end engineering and design), then sharpened in FEED before being released to build and construction. The difference is that the protocols for capturing engineering requirements are much more mature. There are standards for things like steel, and the performance features of assets (weight, throughput, and energy consumption) are well documented. The different engineering disciplines do have their own standards to follow and their respective documentation styles, which contributes to what the industry calls interface issues (e.g., an electrical panel needs to be attached to a physical device, which creates an interface between the electrical team and the structural steel team). Managers in oil and gas rightfully wonder why waterfall, which works reasonably well in engineering, does not work so well with technology.

Compare this to Agile technology development. Built for an age with virtually free computing, networking, and data storage, Agile fixes the time available to deliver some functionality, thus creating the opportunity for more directional feedback. Instead of analysts working with users to agree on the scope in an effort to minimize coding, combined teams of analysts, users, and software developers work collaboratively to deliver a working solution quickly, recognizing that some coding work will be throwaway. Teams work in a series of sprints to refine their understanding of the business problem, and they work transparently together rather than in silos to deliver a minimally viable product (MVP). The interactions between IT and users are transformed, leading to happier users and better solutions. Value, in the form of working functionality, is delivered frequently, in smaller chunks, and not at the end in a large single release.

An important nuance to Agile is the release schedule. New versions of software are released much more frequently than the once-in-six-months model that I was familiar with at Imperial Oil, and not all at the end as with a commissioned gas plant. The coders must work with the team that maintains IT hardware, such as servers and networks, at the same time that analysts, users, and coders collaborate to design the minimally viable product. It makes no sense to have a fast-moving Agile team generating multiple small iterations of a product or solution every few weeks, only to have their efforts hit a slow-moving operations team that can only release software every six months.

Amazon is said to deploy new software to the market 170,000 times per year. Think of the regularity of software releases for your smartphone. That just cannot be done manually. Agile techniques also apply to the team that handles releases, carries out regression testing, automates user tests and integration, executes the actual release, and deployment. Executives ask me why it is that oil and gas companies feel so slow when compared to digital companies, and a big part of the explanation is digital's use of Agile to drive iterative change.

Successful digital teams in oil and gas already use Agile methods for solution delivery. Chances are very high that users don't understand how such innovations could change their business, and digital professionals don't understand oil and gas. Agile teams overcome this limitation. I would have thought that users, with experience only in waterfall methods, would reject Agile out of hand, but that is not so. Users welcome a chance to work with a MVP solution developed in just a few short weeks, rather than waiting months for a more complete but incorrect solution.

Personnel challenges feature prominently in Agile delivery methods. CIOs see the need for a product manager role on Agile teams to drive accountability and decision-making for the solution. The product manager is embedded in the team and must have significant involvement—it's not a part-time, casual job. Teams have to be multidisciplinary and cross-functional, which means collaboration with other managers to find and free up the team. Not everyone can work in an Agile way—it requires that participants unlearn the legacy way they work, which could be a tall order for someone who has worked in the same way for a very long time.

To build momentum using Agile on digital projects, managers seed new digital initiatives with experienced users, analysts, and product managers from previous projects. Managers start projects small (so that value can be quickly delivered) and aim to scale up projects once they reach a high level of value release. Managers have figured out that carefully selecting the team members is key—great team members have to like project work, be ready and able to embrace new ways of working, adapt well with uncertainty, be influential with their peers, and personally grow to take on bigger roles on a digital transformation. Finally, the best of the best see Agile as a way of evolving business culture. Instead of confining Agile to purely technology projects, industry pacesetters use Agile to innovate process areas at the same pace as technology.

LEADING THE CHANGE

The oil and gas industry has not changed much in the past forty years—sure, we have new technologies and have seen multiple waves of mergers and divestitures, the arrival of global shared services, and corporate structure shifts from centralized to decentralized and back again. But there haven't been any fundamental changes that have upended the asset-intense industries to the same degree and with the same impacts as more consumer-oriented industries like retailing, entertainment, and the news media. The industry is fortunate that digital innovations are only now coming to the industrial sector. With strategic vision and thinking, the industry can ride the wave, instead of being crushed by it.

The Origins of Change Resistance

With the possible exception of the fluid, the small, and the fast-moving, organizations tend to resist change. Change is disruptive to safe and effective work patterns and can negatively impact outputs and results. One analyst said that it is clearly written in all middle managers' job descriptions to kill innovation. Oil and gas is perceived to be particularly resistant to change, and indeed, some research suggests that the oil and gas sector is among the most change-resistant of large industry groups.

ORGANIZATIONS ARE LARGELY responsible for killing off their own efforts at innovation. Bob, as a line manager, is told by his boss that if he hits certain performance targets, he will receive a bonus or a promotion. Bob depends on his team to hit these targets. Sue, tasked with studying and implementing some kind of innovation, is told by her boss that if she successfully implements the change, she will receive a bonus or a promotion. Sue realizes that she needs a few weeks' work from three of Bob's team members (who are the business or technical experts) to help with the study and implementation. She meets with Bob to discuss. Bob is sympathetic to Sue, but if he releases his team to work on a digital innovation, he won't hit his targets. The request is turned down.

The culture of oil and gas places strong emphasis on safety, reliability, operational excellence, and environmental sensitivity. Another term for this culture is *risk averse*. Oil and gas is distinctive in how slowly it embraces change, usually waiting for some courageous outfit to try something different and prove that it works over a multiyear time horizon. There's even a name for this behavior: *fast follower*. Change must be nearly perfect in its implementation and effect, and tolerance for ambiguity is low. This risk aversion is applied to all change uniformly, not just those changes that truly demand it. There's a clear preference for "products" that have little feature and implementation uncertainty and very easy adoption. The digital industry is the opposite, with its culture of fail fast, pivot, and rapid cycles of releases.

IN EARLY 2018, I spent a few hours with an entrepreneur engaging on an oil and gas solution he was trying to advance. After some explanation from him, I was able to grasp the business problem he was trying to solve, but for the life of me, I couldn't grasp his solution. There were simply too many unanswerable questions and gaps that I thought created risk. I may be overly risk averse, but if I can't see the value, then I'm pretty sure a real customer—that is, an oil and gas outfit—is going to struggle to see the business case. They dislike being first with anything. Innovation has to be as fully de-risked as possible. Whatever the proposition, it needs to work in their context to get any consideration. They need to see how it will be implemented, not just what it does.

Oil and gas emphasizes technical know-how as a precursor to success. Technical excellence means designing, building, and operating these complex assets to the highest standards of performance, safety, and quality. Quality and risk reviews involve the most experienced technical leads to surface everything that could possibly go wrong. Careers are ruined, and, in the worst cases, lives are placed at risk when technical excellence falters, a critical procedure is not followed, or a poorly thought through change is introduced. Failure is simply not acceptable. Digital innovation places emphasis on upside—what needs to go right, openness, and tolerance—and less on what can go wrong.

A JUNIOR FIELD engineer with an EPC firm was tasked with visiting a company's gas assets and building up an inventory of what was installed. Rather than using the typical paper diagrams and clipboards, he purchased a tablet and downloaded a free app from iTunes that normally would be used to build a home inventory for insurance purposes. He replaced the list of home assets (such as stereo system and microwave oven) with gas assets (such as wellhead, flare, and separator). With a couple of junior colleagues, he then set out to build the inventory. The team photographed the assets, built the inventory, and completed their work in a fraction of the time. The data captured was so accurate and complete that there was no data-entry work for the home-office team. Meanwhile, he encountered heated resistance to his novel approach the entire time—tablets won't work in the field, the junior engineers won't complete the inventory properly, the free app won't be robust enough, the data collected won't be in a format acceptable to the document controllers, and on and on.

The way oil and gas companies are organized reinforces this cautious approach to change. The natural tendency is to group expertise into various technical disciplines, which promotes deep technical excellence but also promotes stovepipe and siloed thinking. Grouping expertise by asset helps to create more integrated solutions for the asset but also asset silos. Individuals who excel in both a technical discipline as well as digital are as rare as unicorns.

Talent shortages are now a factor in introducing change. Following the 2014 price collapse in oil and gas, and with ongoing uncertainty when (or if) prices will return to their previous highs, oil and gas companies embarked on their normal playbook to cope with commodity price downturns. Some 300,000 professionals were let go from the industry globally. There's little appetite in oil and gas to add staff back given the uncertainty, and even less to add skills in areas that are technology-centric rather than in the core competencies. There is little capacity left to explore new solutions, to staff up teams to study how digital innovations could transform work.

On the other hand, low levels of talent turnover, particularly in brownfield plant facilities, tend to create highly cohesive teams but at the expense of nimbleness and agility in the face of change or opportunity. One vendor of artificial intelligence tools told me that the biggest hurdle they encounter is convincing a grizzly veteran who's spent thirty years managing an asset that AI can better his performance at all, let alone by 10 percent or more.

Underpinning the industry's performance is its very strong focus on health, safety, and the environment. All players in the industry, regardless of what segment they are in, track and report incidents, provide high levels of safety training and awareness, and have safety experts on hand to support management in the field. This is completely appropriate. To achieve strong safety performance, managers and supervisors work very hard to encourage the workforce to follow the safe work practices as set out, promoting reliability, consistency, and stability. Embracing digital technology that changes very rapidly is thoroughly counter to this culture.

AT MY THIRD annual performance review at Imperial Oil, at the time Canada's leading integrated oil and gas company, management told me that my career would amount to no more than three promotions over the next thirty years, and that the biggest reason for that was my lack of technical training. Since I was not an engineer, I could not take on a role in any field asset, including upstream, plants, refineries, or terminals. Needless to say, that spelled the end of my career as an employee within the industry...

Managers set very high expectations for economic performance to compensate for the amount of risk that the industry is managing. At the same time, producing assets in rising markets tend to drive very high returns, which tends to drive up the hurdle rates for all investments. This makes sense—anything that yields a return less than that from de-bottlenecking a process, building another retail site, or drilling another well should not be funded, following shareholder value theory. Those expectations translate into very high hurdle rates for innovation, which crowd out new areas, like artificial intelligence, blockchain, and other promising digital solutions. One supermajor has revealed that its hurdle rate for any digital investment must be $1 billion or more, in either improvements in cost or growth in revenues. At that level, not much gets funded.

"Digital is all about people and culture because it's changing the way people work and the tools that they use."

JUDY FAIRBURN, TRAILBLAZER

Different segments in the industry face their own specific performance goals. For example, in the upstream, a very significant pressure on CEOs, driven by Boards and shareholders, is to convert balance sheet assets (or reserves) into cash as prudently and as quickly as possible. Not only do these assets make no money sitting idle on the balance sheet, but the existing producing assets generate less value each year because of reservoir decline curves. This economic pressure gives drilling programs priority over all others. These performance metrics discourage initiatives and investments that could also produce strong returns.

Change resistance is not limited to the fence line. Some industrial technology companies in the supply chain have exploited the industry's risk-averse nature by convincing it of the merits of walled gardens of technology that are proprietary, not open source, with few interconnectivity options—"change, but on our terms." Creative digital solutions struggle to gain traction if they originate outside the industry, beyond the normal supply chain, or need to interact with existing technologies. One supermajor is using a defense contractor to design its next petroleum plant to achieve more openness in plant-automation systems. Companies that supply the military know that warships cannot only accommodate weapons systems from a single supplier—the weapons systems need to adapt over time as new technology is introduced.

Risk #1: Technology-Led Change

The first change challenge to manage is the temptation to view digital innovation as the end, not the means to the end. To quote a gas industry executive from Shell, "People follow process using data enabled by technology." Digital technologies are not the main purpose for oil and gas companies: they are enablers to the greater purpose, which is to find, extract, cleanse, transform, and sell petroleum products safely and efficiently. Digital innovation must be laser-focused on helping achieve this purpose, or it will fail.

The industry thinks long and hard about spending shareholder capital in the pursuit of new hydrocarbon deposits and asset investments, as it should. Geologic uncertainty, technical risk, commodity price movement, available infrastructure, tax policies, royalty regimes, likely productivity of the assets, expected lifetimes, ongoing capital requirements—all of these factors weigh into the decision to allocate capital to an investment. The industry is generally comfortable with modeling these uncertainties and managing them through the capital spend phase. These capital investments tend to dwarf all other kinds of spending, and so receive the lion's share of management attention.

Information-technology investments, on the other hand, are smaller, less familiar to executives, and have less fully appreciated implications. In my experience, technology-led business change is less of a priority, receives less management attention than other investments, is handed to an executive for part-time sponsorship, is not profiled at the Board level, receives limited resourcing in funding and talent, and is not held to a high standard, all of which signal a lack of serious intent.

Poorly structured digital agendas that lack direction, organizational clout, appropriate resourcing, and demanding performance measures will fail to deliver meaningful results.

Risk #2: Orientation to Incremental Change

It is completely reasonable for companies to experiment in one-off digital investments in narrow areas of business to gain insight and understanding of how these technologies behave and evolve. Many oil and gas companies fund a portfolio of smallish experiments and proofs of concept to drive innovation inside the fence or, more narrowly, inside a silo of the business. In my experience, most of these experiments do successfully demonstrate that individual digital technologies can and will benefit the business.

Pursuing a series of individual digital innovations is not going to be enough. Using RPA in the finance function to reduce data-capture costs from field tickets will remove some head count from accounting, but the problem of manual field tickets remains. This is akin to a taxi company publishing an app that allows a customer to get in contact with its call center to book a cab. The call center is still intact, the payment system hasn't changed, the cab driver still doesn't know where to take the customer, and the overall customer experience hasn't changed.

Some portion of innovation investment will need to be redirected towards combinations of technologies working not just in functional areas but beyond the fence line. Some effort should also be directed to understanding how business models in the industry may be disrupted, particularly in the following areas:

- **Data mining:** The potential to separate the analysis of asset data from the ownership of that data and the asset itself—using cloud services to crowdsource analytics for third-party subsurface assets.
- **Vertical integration:** The potential to extend business models beyond current boundaries, to achieve, for example, full inventory management for petroleum products to customer tanks, including vehicles.
- **Derivatives:** The potential to create new financial assets based on highly accurate machine data—using blockchain to record actual environmental emissions (water and air) and enabling trading in actuals rather than estimates.

It is at times of change when opportunities are the greatest. Digital innovation is one of the most powerful, widespread, and fastest change drivers in recent memory, and opportunities to reinvent business practices are at hand. Many industries have learned the hard way that digital innovators can disrupt the status quo in painful ways.

INTEL, A LEADING microchip designer and manufacturer, was presented with the opportunity to develop microchips for a novel device called the smartphone, but its analysis at the time was that the smartphone would not have a market. They passed it up. The music industry, under siege by peer-to-peer music sharing services, decided to sue rather than innovate in music-sharing technology; profitability in the sector collapsed for a decade, and music-streaming companies like Apple and Spotify now dominate. The taxi industry in many cities enjoyed a monopoly of high prices, constrained capacity, poor customer service, and low technical innovation, until Uber appeared and thoroughly destroyed its business model.

Risk #3: Organizational Misalignment

Another risk that could squander digital investment is "inside the fence" organizational misalignment. For many digital investments to achieve their full potential, business functions outside those managed by the main sponsor need to get on board.

For example, as I have described above, one of Australia's coal seam gas companies experimented with using a collaboration tool in its field operations to convert the traditional well delivery from a discrete project effort to more of a manufacturing process. This initiative, called the Gas Factory, sat with the operations team, who scheduled well delivery, contracted with the services companies, and engaged with the landowners. Progress was stellar—well delivery started out at four to five wells per month, and in time ops began to deliver thirty wells per month, one well a day, seven days a week. As in virtually all upstream companies, finance owned the authorization for expenditure (AFE) process, which stipulated that all wells required Board approval. Aside from the fact that the Board only met infrequently, the finance process required at least a week per well of financial review, analysis, and modeling. The finance function had not bought into the idea of mimicking manufacturing—they were hired into an oil and gas company. They were unable and, in some cases, unwilling to change their whole process to match operations' speed and function.

The most impactful digital initiatives will involve the whole organization—not solely its components and not merely inside the fence. Innovations that collect data directly from sensors in the field or on assets or from suppliers, that process that data in a fully automated fashion, that apply machine intelligence to assist with interpretation, that turn the data into incremental value through sharing will drive outsized value. Leaders need to recognize when organization power structures might block successful innovation.

Risk #4: Short-Term Thinking

Maintaining funding for digital innovation across planning horizons takes organizational will. A digital innovation is unlikely to be the best marginal investment in oil and gas at all times of the economic cycle. Consider just the simple math of a producer of 100,000 barrels per day. Assume the price of oil moves from $45 to $50. That's an addition $182 million in revenue in a year and no increase in cost. For most companies, the smartest economic decision when prices are rising is to direct resources towards incremental production, which may mean cutting digital innovation.

Effectively, the only time digital is the best marginal investment is when production costs are well above the recovered price (the net back is zero or negative), supplier costs have been slashed sufficiently to trigger bankruptcies, capital spend is just keeping the lights on, and the workforce has been trimmed to skeletal levels.

Oil and gas companies with a history of short-term thinking on innovation in general, and digital in particular, are unlikely to achieve much benefit with a cyclical pattern of investment followed by disinvestment. Talent will quickly figure out that their efforts are unlikely to be successful and will behave accordingly.

Tactics to Promote Digital Adoption

I draw a distinction between change management—the discipline of introducing people and process change into a business—and the management of change (MoC), the process for changing an industrial, chemical, or mechanical process in a plant setting. In an office setting, digital adoption is mostly about change management, but in a plant setting, where wireless digital sensors may be installed to provide improved data and insight, digital adoption may be both change management and MoC.

There are plenty of great reference works and many available advisors to help with crafting and executing change-management programs, and the MoC process in industrial settings is already deeply entrenched. But there are a few key tactics to promote digital adoption for oil and gas that are worth mentioning.

THE CIO OF a large integrated oil and gas company is changing its internal language of digital innovation from "fail fast" to "learn fast" to shift the culture of the organization. The fear of failure is deeply ingrained in an industry that has to deal with the occasional catastrophic incident.

Since the high hurdle rates in oil and gas tend to block innovation, change the hurdle rate for digital projects to encourage some trials. Set aside specific budgets for digital projects that cannot be allocated to other ventures. Force specific targets for digital innovation, such as reduction in cost or improvement in productivity, and hold managers accountable for delivery.

Digital innovations may be applied in many aspects of an oil and gas operation, but receptivity is probably highest where the competitive gap is the greatest. In the plant setting, use the Solomon benchmarks to identify performance gaps. Focus on areas that manage or improve carbon dioxide and methane emissions, or improve the cost of operations.

For the committed, set up an executive role responsible for digital and give them real clout to raise digital awareness, influence digital investments, resolve data issues, design organizational approaches to digital innovation, embed digital thinking in the business, and champion success. Some organizations have created the role of chief digital officer to drive ownership and results.

Take a page from Woodside (discussed on pages 43–45) and create a small team focused on driving experiments and value creation using digital. (In Woodside's case, the team is focused on artificial intelligence.) Leverage the ecosystem of vendors and consultants to build that capability quickly. Equip the team with the tools to do the job.

Try launching a number of small digital trials in a portfolio to see what actually works. I would scope the trials and technologies that do not require a change in the physical world of steel and cement, to avoid triggering a detailed management of change process. Further, I would time-box the trials—they don't need to run as long as physical-world trials to understand their impact. The trials would be "Agile" in that they run as a series of sprints to produce measurable outcomes quickly.

IMPLEMENTING RETAIL SYSTEMS

Case Study #9

A few years ago, I was working with a downstream oil company that was, by its own admission, well behind the times. The cash registers at their retail stations were so obsolete that they resorted to scavenging parts from underused ones. Credit card authorization took a full thirty seconds because the equipment was so old. Debit cards could not be processed because the cash registers' software didn't recognize that kind of card.

The competitors had long since introduced a crazy new innovation called "pay at the pump," something we all now take for granted (unless, perhaps, if you live in parts of Oregon and New Jersey). But for my client, with its obsolete systems, pay at the pump was not feasible. It was becoming a matter of competitive survival—unless the company solved these issues, customers would continue to drift away.

The company decided to embark on a sweeping project to overhaul the entire customer experience, both inside the retail site and at the pumps. The scope was audacious at the time—all new signage, colors, and branding; new cabinetry and cash registers with scanning capability; and new software. One goal was to sort out pay-at-the-pump technology and make it work in concert with the new cash registers.

My team and I faced all sorts of challenges. The fuel pumps were too expensive to replace—they were mounted on concrete in the forecourt—so we had to work within that constraint. Fortunately, they included some removable front panels that could be swapped out for card readers. Some sites didn't have good network coverage, and they needed to trial a satellite connection. The number of different credit and debit cards to process numbered in the hundreds. And for the first time, the fuel pumps with their new pin pads would need to connect to the cash register system inside. Trial and error wasn't really an option, since the physical world of steel, cement, and buildings was involved, and it wasn't affordable to visit all the sites more than once.

Think of all the players involved in making this system work—the inside store supervisor and counter service staff, the motoring public, the accounting team, the bank technology teams, and the telecoms service (satellite and terrestrial). Then there's the hardware—fuel pumps from Gilbarco, the CRIND technology, pump controllers, the new cash registers, modems, and scanners.

THE CONFERENCE ROOM PILOT

To deliver pay at the pump for the customer (not just once, but at 750 stations, eight pumps per station, in two countries), the team needed to make dead sure it would absolutely work in all weather conditions, in multiple languages, and in multiple currencies. It needed to be easily installed. Customers needed to be able to use it without training. And it needed to hit a sub-second response time. No one was going to wait thirty seconds in the rain or cold for a credit card authorization.

Developing and testing the solution was complicated by the fact that the team couldn't exactly take over a convenience store. The solution was to commandeer a conference room at the home office, within which the team recreated the entire pay-at-the-pump experience. The room featured a working fuel pump that dispensed imaginary fuel from the nozzle. The setup included cash registers, scanners, modems, satellite connections, and test debit and credit cards. Desks and seating mimicked the places of work for the cashier, the shift supervisor, the back office, the bank, computer support, help desk, and other key roles.

The team unrolled long sheets of paper and pinned them to the walls and, using Post-it notes and markers, mapped out all the various processes that would happen—everything from a working purchase, a stolen card, over-limit cards, theft from drive-offs, cold weather, high-volume activity, a network going down, interrupted transactions, and mixed inside/outside sales. That paper included the capital cost and budget, flow charts, checklists, problems to fix, project plans, documentation, performance targets and measures, and other project data. The process flows highlighted what integrations were needed for the existing systems including finance, human resources, and store inventory. As the team worked the solution, the walls were continually updated. In time, we perfected the solution and were able to implement it successfully across the fuel stations.

The challenge I see with digital innovators and entrepreneurs today in the business-to-business area is that they tend to take too narrow a view of what their solution is going to deliver and how it is going to work. They don't fully think through the business model that they are building, including how to make money, how to stimulate rapid take-up, and how to enable other market participants to collaborate. This is perhaps admirable in maintaining focus, but I fear that it does not result in a minimum viable product that an oil and gas business can embrace. In my "pay-at-the-pump" example, the MVP had to incorporate a pretty complete range of features to be minimally viable, in addition to working with banking (which is rigid in how it works) and other suppliers.

B2B PRODUCT DEVELOPMENT DONE RIGHT

If today's digital innovator, either as an internal change agent or as an external solution company, is going to succeed in selling its solutions to a demanding industrial audience, it needs to take a page from yesterday's digital playbook and bring more complete solutions to the table.

- A digital solution that only reflects one end-user experience is simply too narrow. Think about all the possible personnel in different roles and jobs inside a corporation that could be affected by your solution. At the very minimum, you'll encounter financial controllers, accountants, and engineering services.
- Your digital solution must fit within some business context. Try to draw out the process so that you can at least show the inputs and outputs and where you anticipate integration (probably into some existing company system). More personnel will start to show up here. Think about the use cases and scenarios, particularly the most extreme ones.
- Unless your solution is purely in the cloud, like an AI algorithm, it will have a hardware component. If so, you need to show how it integrates with at least some of the hardware category leaders and how the solution is extendible to other providers. It goes without saying that you need to fit in with all the requisite user hardware like tablets, smartphones, and browsers.

- It's a good bet that your digital solution takes a fresh look at data, either by creating new data (such as GPS location data, visual data, distributed ledger data, or tokens) or aggregating data (in the cloud or via analytics). Who uses this data (thereby involving a new person)? If your solution needs data that might exist already, where does it come from? More people. It's a pretty good bet you'll need to integrate with ERP systems like SAP and Oracle.
- The oil and gas industry is very safety and environmentally conscious. You need to be able to show that your solution does no harm to people, the environment, or equipment, and ideally that it improves safety conditions. How will you respond to questions about cyber security, such as hackers, and other digital vulnerabilities?
- Business will want to understand how your solution will behave under varying conditions. This is particularly true for blockchain solutions that incorporate some kind of token with some kind of valuation. Imagine a scenario where the barrels of oil in a reservoir are sold forward, and each barrel is recorded on a blockchain and represented as a token. You need to be able to show, on a model, how the tokens are created, transferred, sold, traded, swapped for cash, valued, and destroyed. Is there a two-sided market that is being unlocked? How does your solution behave at different market volumes, commodity prices, currency values, and interest rates?
- My "pay-at-the-pump" example made money by raising traffic volumes and increasing throughout (all those thirty seconds add up). But how do you make money with your solution, both for you and for your customer? This is particularly true for blockchain solutions that create value via tokens.

KEY MESSAGES

Deciding how to get started with digital innovation is a universal challenge for this industry. Operations are massive, complex, and spread out, and transitioning to digital has to be done while running the business. But it is crucial to get started somewhere.

Key takeaways from this section:

1. The risks and challenges of digital transformation are common to many industries.
2. Digital is a significant change for an industry focused on safe, secure, and reliable processes. No matter which group manages the initiative, change must be driven by the Board and executives.
3. Getting started is far more critical than starting in the exact right place.
4. Brownfield assets will adapt to digital much more slowly, if at all, than greenfield assets will.
5. Data integrity will underlie all digital successes (or failures).
6. The cyber security risks facing field assets equipped with sensors pose a considerable challenge to oil and gas.
7. Digital technology is actually the easy part. Managing change in the workplace is the hard part. Human workers will need to be agile and embrace learning and continuous self-improvement.

NOSE IN, HANDS OFF:
The Board's Role in Moving Forward

I LIKE THIS PHRASE—"NOSE in, hands off"—to describe the Board's role. The Board should take an active interest in the activities of the company but should let managers manage. When it comes to digital innovation, this means getting smart on digital, framing a perspective on how it is evolving, setting out strategies for reacting to and taking advantage of developments, creating new structures to tap into digital capabilities, and setting goals and targets. This chapter sets out how the Board might incorporate a digital element to its supervisory agenda.

GETTING SMART ON DIGITAL

Boards need help to understand and endorse digital strategies and investments. Blockchain, augmented reality, the Internet of Things, additive manufacturing, machine learning, crowdsourcing, social media—these are all modern inventions. Board selection committees sensibly seek experience for their ranks, and the digital industry is too young to have spawned enough mature, experienced, Board-quality hands in anything like the number needed to meet demand. Many Board members tell me that as a group they lack much direct and personal experience in digital. Many dwell on the long and difficult efforts to adopt ERP and other enterprise tools, worrying that other digital innovation will be more of the same.

Boards will find it a challenge to properly address digital issues without more expertise. Boards should be asking management to invest in getting digital expertise inside the company. That expertise could be in the form of a small team, whose task is to help the company and Board understand and anticipate the digital shifts in the economy.

Boards should also try to add some digital expertise to their ranks. The biggest oil and gas companies have already added digital depth to their Boards. Retired Google executives may be both rare and expensive, but there are bound to be serving executives in the more mature technology companies who could be valuable (think Cisco, IBM, or Telco). Anadarko has a thirty-five-year-old CEO of a digital startup on its Board. If not full Board members, there is certainly room to set up digital advisory councils and board committees.

If digital is struggling to get into oil and gas, then take oil and gas to the center of digital action. By this I mean Boards could pay a visit to a campus of a digital giant. Yes, digital companies are a bit secretive, and they all seem to have initiatives intended to disrupt transportation fuels, but in general they want new markets, too, and they would value getting to know such a revenue-rich industry. Many have announced industry verticals (Apple) or are already widely deployed in oil and gas (Microsoft) and they would see value in hearing from the customer.

There are lots of incubators and accelerators that are constantly looking for funding and participation. Some are freestanding, others are part of a university or community college system. Membership is generally modest ($50,000) to get a front-row seat to the developing action. I would conduct a scan of the incubators and accelerators and confirm who might be looking for sponsorship in exchange for some influence over which technologies and companies the incubator nurtures. And note that the best accelerators for a company might be far from its hometown—digital innovation can happen anywhere.

Boards should be challenging their own ranks about why they are not actively seeking advisory roles on digital startups. Startups lack business experience—most are headed by a founder with an idea, and they need all kinds of help of just the sort that a Board member can offer. I'm seeing this firsthand with the community that I advise. Digital startups need insight into how to sell to oil and gas, how to overcome adoption inertia, and how to price their innovations.

Today's university student has fantastic exposure to all things digital and the know-how to build prototypes and to prove their ideas (recall that today's digital giants generally started with a fresh grad and some breakout ideas). Boards should invite the local business school students and campus clubs into a pitch-off. Boards would benefit from seeing what the true frontline is working on (and the implications for attracting and retaining next generation talent) and students always appreciate a free meal.

THE WORLD'S MOST aggressive digital country by far is Estonia. A new mandate made it illegal for a government department to ask its citizens for the same information twice. The country had to agree on a national data architecture to achieve this goal. It attracts more virtual citizens than any other in Europe. Boards might wish to hold a meeting there and see about meetings with the country's CIO to understand how it is transitioning to a digital future. Estonia's insights will fundamentally challenge many Western orthodoxies about commerce, governance, accountability, and business models.

Finally, given Boards' digital deficit, some kind of continuous learning and awareness program would help Boards execute their role, including regular briefings on the latest developments and the impacts on the industry. Once a year is clearly not enough.

RAISING BOARD AWARENESS

Some skeptics in management and on boards say, "Digital can't really happen in oil and gas. The industry is too regulated, too fragmented, too business-to-business. Operations are 24/7 and can't be taken offline, except in turnarounds or emergencies. Shareholder returns are still best served by doing what we have always done." I've heard these comments personally at conferences and in workshops across the industry.

But digital topics are now rising to the Board for discussion. Requests for support for a cloud-computing initiative may have made its way up the line for approval. Or the company has been hit with some ransomware cyber-attack. Or a Board member sitting on a financial services company may be wondering why they're not seeing the same level of interest in digital from the resources sector. Or there are some early digital results that have fallen short or wildly exceeded expectations, which triggers introspection.

BUSINESS MODEL DISRUPTION is occurring with more frequency and to more industries because Silicon Valley understands how to configure businesses and business models so that they can scale up very rapidly. They accomplish this feat by making adoption cycles much shorter. A device in every pocket is a start, but couple that to a software solution so easy to use it requires no training, is distributed through an app store to a global market, runs on cloud computing, plays like a game so that it's mildly addictive, is funded using venture capital, and is promoted through social media. These mechanisms are available to all companies seeking to disrupt the competition.

Boards of oil and gas companies seek tangible and practical actions to address digital change, if for no other reason than to respond to market pressures. Most, if not all, publicly traded oil and gas companies have some kind of stock-option scheme or share-distribution program for their employees and managers. My sense is that to put some froth into the share price of an oil and gas company, it needs to have a story to tell capital markets about how it is dealing with technology-driven change. This must start with conversation about particular topics at the Board level, which has the final say on where companies invest.

Demand Destruction

Boards study questions such as:

- How are we preparing for eventual decline in demand for fossil fuels?
- How will digital innovations such as automation, AI, and additive manufacturing accelerate the decline in demand?

Boards are concerned with the underlying demand drivers for fossil fuels and the preparations underway to manage this period of fundamental demand shift. A quarter of each barrel of oil is converted to gasoline for the transportation sector. But for the first time in a very long time, anyone can take a test drive in a thoroughly viable, affordable, and more environmentally responsible alternative to fossil-fuel personal transportation.

Boards note that these new vehicles from Tesla, BMW, and other leaders are incredibly appealing, capable of achieving ludicrous speeds, are ultra-quiet, and feature high-tech interiors. They function more like what next generation-buyers think of as a technology—as easy to use as smartphones. Plug it in for a few hours, download some updates, and you're good to go. Indeed, they're more like pure technologies and could behave according to Moore's law. The oil and gas industry has yet to come to grips with this impending change to personal transportation. The competition in the automotive sector is going to redefine personal transportation, changing consumer behavior to favor new technologies.

In fact, the next generations of buyers might not even see car ownership as a personal imperative. My kids see driving as a frustrating time-suck when they could be watching videos, surfing the web, or texting friends. Smartphones come first in their lives, not cars, and the online experience just keeps getting better and better, compared to driving a car, which is just getting worse and worse (think traffic jams, noise pollution, parking hassles, and expensive repairs).

Boards are intrigued by the possibilities posed by vehicle sharing and the breathless hype around self-driving cars. They discuss how these fundamental behavior shifts will translate into permanent demand shifts. Researchers have modeled, for instance, how many taxis might be required to supply all of New York City's surface transportation needs once these technologies are solidly in place, and it's just a few thousand, compared to over 14,000 licensed taxi medallions, thousands of other hire cars such as limousines, and all the private vehicles roaming its streets in 2018.

Of course, there are a billion cars in the world, and global annual manufacturing capability is 72 million combustion-engine cars and light trucks, or thereabouts. It would take thirty-three years to phase out the existing fleet. I'm ignoring the learning-curve effects associated with manufacturing, but think for a minute what happens when personal transportation as a technology starts hyper-improving through Moore's law. The head start that fossil fuels have could evaporate in a hurry.

Boards who move their meetings to different global locations discover that the early adopters are hidden from easy view. They're likely in China. While North America might raise the bar for accepting self-driving cars on shared city roads, the Chinese might simply mandate it. And fossil-fuel growth is predicated on continuous Chinese demand growth.

Shifting Cost Structures

Boards study questions such as:

- What is the impact on the business from rising costs for water and emissions and changing economics for energy inputs?
- How do we leverage the Internet of Things, robotics, and blockchain to optimize our business?

Boards challenge management to outline how the business intends to reposition itself as a low-cost, zero-carbon, and water-neutral fuel supplier. As the economics of energy, emissions, and water shift, business models need to adapt, and digital technologies can help with that.

Consider the economics of energy. The oil and gas sector, from extraction of crude oil to retailing of petroleum to customers, consumes a significant amount of energy, including diesel and natural gas. But input energies are rapidly evolving with the rise of biofuels, wind, and solar, supported by batteries and smart grids for distribution. These new energy sources are falling dramatically in price and are the energy of choice for the rising challengers in personal transportation. In the main, the fossil fuel industry is highly reliant on its own product, and that product is becoming a high-cost fuel. Boards are starting to ask which executive is in charge of energy inputs (to match the executive structure in traditional areas such as supply chain, HR, and production).

Next is the economics of emissions, or greenhouse gases (GHGs) such as carbon dioxide and methane. Under the terms of the Paris Agreement, emissions are going to be highly scrutinized in Europe, China, India, and Asia. Countries are introducing tax policies (carbon taxes, or cap-and-trade schemes) to create the economic incentives for action on emissions. Boards know that the oil and gas business sector is a significant source of GHG emissions after coal, through the tailpipe and flare stack, but also through rogue emissions from equipment. Many in the industry are embracing a carbon tax, in part to level the playing field with coal, but also to create the right economic incentives to tackle GHGs.

The trajectory is clear: fuels that emit non-degrading or non-useful gases to the atmosphere are going be sanctioned with either costs, penalties, or both. Boards are actively reviewing the impacts of climate regulation on oil and gas, with keen interest in how business is measuring and managing emissions, a task for which digital is rather well suited. Blockchain, the Internet of Things, and cloud computing will help enable tracking and trading of carbon credits.

The final economic shift relates to the use of water and the rise of water in the social agenda. Water features very prominently in oil and gas operations: as a fluid in fracking, as a lubricating medium for drilling muds, and as a treatment environment in settling ponds for oil sands. Water is both a waste by-product of the coal seam gas industry, where it may require extensive desalination and decontamination treatment, and a key input as steam for bitumen extraction and thermal well operations.

Society is becoming increasingly alarmed about water usage in oil and gas. Despite the overwhelming evidence that fracking is an environmentally safe technique, skeptics believe that fracking contaminates drinking water. The number of marine accidents involving oil spills has fallen to record lows, much the same way as airplane crashes have been reduced, but activists are resolute in opposing crude oil shipping. Proven methods of safely moving hydrocarbons via pipeline are being throttled by the environmental movement. Governments are capping resource extraction because of stress on water resources, which is stranding those resources.

The Pace of Change

Boards study questions such as:

- Are we transforming our business fast enough and deeply enough to deal with competitive pressures?
- Are we adequately considering new digitally enabled business models as both competitors and opportunities?

Boards challenge companies to adapt work processes and to report if the work has materially changed in response to these environmental developments. Other industries have abandoned traditional ideas about what is core to their business and what can be more profitably sourced. Digital solutions are becoming pervasive in finance with online banking apps and in retail with online shopping. New business models have disrupted formerly stable industries, including power generation, entertainment, and the media. Industrial concerns ask if some long-standing and deeply rooted organizational structures (such as the historic separation of operational controlling systems in plants from the commercial systems supporting supply chain, finance, and talent) are a liability in a world with cyber-security concerns.

Oil and gas has historically followed a one-dimensional approach to adaptation, which is to do more of the same with less, instead of doing different things differently. During a downturn, the industry normally cuts off innovation across the business (in the form of funding, resources, and management attention), leaving it poorly positioned for the future. This may no longer be the right strategy, given these big conversations on the shifting economics of energy, emissions, and water; the growing alternative fuel options for consumers; and the impacts of business-model shifts.

Boards recognize that the oil and gas industry is very innovative, but it changes very slowly and narrowly. They now challenge why the limited available capital for innovation only goes to extracting more resources more cheaply and to process issues like improving fracking and drilling. They press to see that other innovation levers are not ignored. Boards are familiar with the arguments in favor of the go-slow approach to change (the industry is dangerous, highly regulated, etc.), but because of exposure to other industries, Boards challenge whether the slow pace is an excuse. Consider how quickly the automotive industry is experimenting with autonomous vehicles on city roads.

Social Acceptance

Boards study questions such as:

- Are we maintaining our social acceptance through safe operations?
- Are we adequately prepared for an inevitable cyber event?

Annual general meetings for oil and gas companies have taken on a siege mentality. The industry faces challenges from all quarters, from investors asking about the sustainability of the industry and the security of their investment, to activists blocking new developments at every turn, and to regulators upping the compliance burden. The social acceptability of the industry is neutral at best.

Boards insist that safety and security of operations are not negotiable. Incidents such as spills, explosions, fires, and fatalities reflect badly on the industry as a whole, but are particularly problematic for individual companies. Tremendous energy and shareholder capital are spent training the workforce, securing operational facilities, improving performance, and reducing the negative impacts of the company on the environment and society.

Boards are acutely aware of the threat of cyber activities. The media highlights cyber events regularly, and Board training sessions now delve deeply into how cyber events unfold, and the consequences for the Board, management, and employees. The critical vulnerability for the industry is in operating infrastructure. All the remote sensors and devices that connect physical infrastructure to the company are beyond the secure fence line and are vulnerable to cyber activity. As has already happened to energy infrastructure elsewhere, it is inevitable that cyber criminals or their bots will target oil and gas infrastructure (pipelines, tanks, pumps, valves, and power) and unleash chaos. Boards want to know that management is anticipating cyber events and that response plans are tested, in place, and ready to go.

Attracting Talent

Boards study questions such as:

- Are we hiring and retaining the best talent for our business?
- How are we transitioning our talent models to attract, hire, motivate, and retain a digital workforce?

Boards have always taken an interest in the talent of the industry and, more recently, with making the sector more attractive to a diverse workforce. The industry has not faced talent shortages in many decades and has been a talent magnet in technical fields. However, the most recent bust, in 2014, was particularly tough on talent—some 300,000 jobs were eliminated, and industry executives openly speculate that many of these jobs will never return. Job seekers look skeptically at an industry with low job security, and unfortunately the industry telegraphs that its jobs are vulnerable to economic disruption.

Employees have also been schooled in the discipline of the market through the downturn. They know they are dispensable and will be looking to future-proof their own careers. Employers that are slow to embrace digital in their companies are also laying the groundwork for talent flight. That talent will have many new options.

The industry needs to confront fresh talent challenges—just as the energy transition gets underway, young talent may conclude that there is no long-term future in fossil fuels. Human capital with the kind of skills needed to help the industry take advantage of digital solutions is highly sought after by other industries with better long-term prospects. Reward models that historically favor geologic and engineering know-how may disfavor the kinds of digital skills that the industry needs in the future. Data scientists, algorithm developers, and robot designers may find the industry too uninspired and too risk averse, compared to other sectors that are also undergoing transformation.

Boards want to understand what management is doing to transition its human capital systems and models. How will companies attract these new digital skills to the industry, in the face of competition from so many other sectors? How will the company create the right environment to motivate a more digitally savvy workforce? The career ladder in oil and gas provides ample upward opportunity for engineering and technical skills, but IT skills are not normally seen as core to the industry. How will talent development and growth models in the industry flex to allow new workers to flourish in the industry?

Capital Allocation

Boards study questions such as:

- Are we investing appropriately for the future?
- How much should we be investing specifically in digital innovation?

Boards need to balance the investment portfolio to provide for current needs as well as for future growth. Some companies will invest more than others because they stand to gain more. For example, for fast-cycle capital (that is, small capital investments that are relatively quickly put into production), digital enables growth. An unconventional upstream player that drills $10 million shale wells is ideally positioned to apply digital techniques to the well-delivery life cycle and transfer lessons and benefits to subsequent wells. I expect those companies to disproportionately benefit from digital.

For long-life capital, digital enables survival. Once the capital is spent, particularly on the plant and equipment, it needs to hit high availability targets, run at high utilization, deliver high throughput, and produce on-spec output. Equipment providers foresee digital improving all of these critical performance curves in dramatic fashion. Converting kit to be more digitally enabled or installing only the latest digitally smart kit will be key.

For equipment providers, digital is becoming table stakes. Some equipment houses are going to see digital as offering a competitive advantage and will so convert. Faced with the option to purchase either dumb or smart gear, some procurement teams will favor the digital equipment to have better innovation ability in the future.

There are clear benefits to investing a little in digital today. First is to capture the benefits from digital developments that have been de-risked. A good example is cloud computing. The cloud sector is now led by a few very large organizations (Amazon, Google, and Microsoft) and most new software is being built for the cloud. In fact, business may risk stranding its legacy software investment by continuing to invest in on-premises software solutions and proprietary data centers.

Second is to gain insight into the future. Digital technologies are evolving very quickly, and it's not obvious where these technologies will go, nor what benefits they might ultimately deliver. A good approach is to work with an ecosystem of digital technology outfits (the majority of which know little about oil and gas) and run small pilots and proofs of concept to see what the benefits are likely to be. For example, it was not initially clear that cars could benefit from being direct participants on blockchain, but after a short trial, the automakers became convinced.

Third is to build talent and capability. Learning how to move a digital innovation from concept to operations is simply not a competence that is widely available inside most oil and gas companies. It will take time to build that ability to the point that it can take small projects from start to finish, then progressively larger projects, and to train up other digital change agents.

Do-Nothing Consequences

Boards study questions such as:

- Is a do-nothing strategy viable?
- What are the consequences if we wait and see how digital impacts the industry?

Most Boards want to understand the downside risk of any investment, including digital. It's a prudent question when you're responsible for shareholders' money. One risk is a compromised market position. Market analysts are starting to ask companies to sketch out how they are responding to the digital wave and what investments are being made. They are comparing companies and gauging competitive position. Discounts will be applied to companies that are harvesting and not investing in the digital future.

A second risk is the widening digital gap between how business operates today, how it could operate in a digital future, and the increasing challenge to cross over that gap. Eventually, the gap may be so wide that it sparks a crisis, either from a complete lack of capital to invest in the upstream or a production cost out of proportion with peers in the midstream, or an eroded market in the downstream.

A third risk is that some competitor emerges from seemingly out of nowhere with an innovative business model. There are plenty of examples in other industries, and emerging evidence that the petroleum industry, or least parts of it, may be particularly vulnerable to new entrants.

Moving Forward

These conversations help Boards work with management to secure the future and minimize any value loss. With digital innovations impacting all parts of the economy at the same time but in unequal measure, Boards and management should be giving some thought to the many ways this uncertain future might unfold. Scenario planning is a useful technique that helps frame these uncertainties and provides some directional clarity for investments. Boards should then be asking management for a digital strategy, that accounts for these scenarios, the company's economic position, how it stacks up to peers, and how the company will close performance gaps or capture real tangible growth from digital investments.

SCENARIOS OF THE FUTURE

After raising their digital awareness, and having some deep conversation on how digital may impact the industry, Boards should then give thought to what a more digitally enabled future might look like. Many companies have what I describe as a bottom-up digital program underway—small investments, sprinkled throughout the organization, low funding levels, and unintegrated pilots that are generally not working towards a broader vision. Usually there isn't even a vision—just an incremental business plan. This can work, but it risks moving too slowly relative to a competitor with a more determined approach, and it is completely vulnerable to the sudden arrival of a new digital-driven business model. More sophisticated organizations take a top-down approach that gives thought to how the future may play out in the form of various plausible scenarios. Scenarios are, of course, precisely wrong, but can be directionally insightful and, with some work and analysis, can be very valuable for Boards and management to frame direction, select priorities, and allocate investments.

BUSINESS-MODEL CHANGES SEEMINGLY come from nowhere to catch industry incumbents off-guard. For example, Kodak had its eye on digital cameras, but the real threat came from camera phones. Paper maps have all but disappeared with the development of GPS. The Yellow Pages were not searching for search engines. The DVD industry was worried more about LaserDiscs than video streaming.

One technique that helps shape future scenarios is to consider the trend lines that look very hard to stop (i.e., there's a solid enough fact base to suggest they'll continue) and the big uncertainties for which no one has an answer. Map these out and see how they inform possible, but uncertain, futures. There are many trends worth consideration in oil and gas, but here are five that seem unstoppable to me.

SCENARIO-PLANNING WAS PRACTICALLY invented by the oil and gas industry. Every three years or so, Shell issues its long-term scenarios for energy markets and is widely anticipated, not just by the fast followers in oil and gas but by investors, governments, suppliers, and regulators.

1. **Demographic shifts.** Demographics favor growth originating in Asia to match the fast-growing populations and their move into the middle classes. Their population will want more of the consumer goods that use petroleum energy (cars, industry, or chemicals).

2. **Technology advances.** Technology is advancing at a breathtaking pace, without regard to the kinds of traditional constraints to innovation, such as capital, capability, and organizational capacity. Technology enables oil and gas to expand resources, at the same time as it destroys demand.

3. **Climate change.** Developed economies have proposed new carbon taxes, fuel-consumption taxes, fuel-efficiency targets, and even bans on combustion engines.

4. **Transportation shifts.** Transportation, as a big consumer of energy, is also shifting. The six top automakers, which account for 50 percent of all cars and trucks, have all announced plans to electrify their vehicles in the next five to ten years. The heavy truck sector and bus sector have started to convert to new drivetrain technology.

5. **Oil market shifts.** North America will soon be self-sufficient in oil, which will cause oil trading patterns to change. Digital innovation will shift the recoveries from shale to match those of conventional, unlocking yet more hydrocarbons, and it may also bring about demand destruction.

Surface Critical Uncertainties

In as much as the trend lines have some broad predictability, life is uncertain. The global economy, for example, could continue to benefit from the globalization efforts of the past thirty years or could retrench or stagnate due to rising nationalism and economic protectionism. Governments could put policies in place to promote greater digital adoption or could block digital advancement through bans, regulation, or nationalization.

- **Uncertainty 1: Global economic performance.** Growing economies consume greater quantities of energy to drive growth, whereas unstable environments cause price inflation and supply pressures. Will we see a return to the stable geopolitical environment of the past fifteen years with its strong globalization flavor, or will we see the continued chronic and increasingly confrontational stagnation that has plagued much of Europe since 2008?

- **Uncertainty 2: Government digital policy.** Governments are behind in addressing digital shifts. We see this play out in markets where Bitcoin, Uber, and Airbnb attempt market entry and run headlong into regulations and institutions that favor incumbents. Data is the lifeblood of digital, but will policy makers lighten restrictive privacy and national governance rules around data to enable innovation? Will trans-border data rules be relaxed to allow innovative solutions in artificial intelligence and machine learning to transform whole sectors of economic life? Some nation states (Singapore, Dubai, and Estonia) have embraced digital change, but will the others?

Map Them Out

We can create four scenarios of the future if we map out a stagnation versus globalization world against a digital versus analog world, and imagine how the five unstoppable trend lines might play out. Each of these scenarios has implications for the oil and gas industry and its relationship to the digital sector. I've nicknamed these quadrants after movies of the past two decades, partially in jest, but also because some cinematic efforts, like *Star Wars* or *Avatar*, try to paint a portrait of the future and give life to these visions.

Example: Digital O&G Scenarios

	STAGNATION	GLOBALIZATION
DIGITAL	**BLADE RUNNER** Technologically advanced dystopia, with uneven global living standards, international isolation.	**THE JETSONS** Rapid digital adoption and global access create new business models.
ANALOG	**THE FLINTSTONES** A move backward technologically to analog and manual processes. Governments invest at home.	**BACK TO THE FUTURE** Strong globalization, but digital falters with governments reluctant to embrace changes

ANALOG VS. DIGITAL

STAGNATION VS. GLOBALIZATION

THE JETSONS

In the upper right quadrant is a strong digital world with a continued emphasis on globalization. It reminds me of the cartoon series *The Jetsons*, with its futuristic carefree lifestyle and robotic assistants. With the world on a global growth agenda and the rapid adoption of digital technologies, the demand for energy continues to grow unabated, particularly in developing and emerging economies; and digitization of the economy facilitates renewables and electrification, which are most amenable to digital enhancement. Digital, enabled by eager governments, rapidly transforms significant segments of the global economy—unlocking new business models, accelerating trade, and unleashing demand growth. Fossil fuels lose ground to electrons. The application of digital technology to the industry accelerates, particularly in shale resources, which have low recovery rates today but could dramatically improve.

BACK TO THE FUTURE

In the bottom right, *Back to the Future*, strong globalization continues, but digital change flounders. We have more of the recent same, but with the brakes on digital. With the world on a global growth agenda, but with governments resistant to adapt to digital change, digital transformation of energy demand centers such as transportation, industry, and buildings stall, with entrenched players blocking innovation. Oil and gas companies apply digital thinking inside the fence to improve productivity and lower cost. Demand for fossil fuels remains strong in our analog world, with 1990s business models. Traditional fuels maintain their dominance as institutions, tax policies, and regulations block digital efforts. The climate suffers as the growing population adopts the energy-intense lifestyle of the West.

THE FLINTSTONES

In the bottom left, economies stagnate amid retrenching global economies. Call it *The Flintstones*, backward, manual, and unautomated. Low global growth favors established energy sources like fossil fuels, particularly from low-cost producers with large established resource bases (OPEC) to the detriment of nonconventional sources (shale and oil sands). With taxes under pressure, rising unemployment, little funding available for innovation, and carbon still on the agenda, governments hunker down. Restrictive rules on digital accelerate the rise of China, whose government system is more directive. A few early adopters of digital look like winners, but jobs at home count for more.

BLADE RUNNER

In the top left, economies stagnate as digital takes off. Low global growth, and a resulting low demand for energy, plays into a job-scarce digital world, with extreme wealth disparities, high unemployment, low pay, and low tax revenues. Governments have no option but to play along with cryptocurrencies, autonomous trucking, and robo-farms, or risk being globally uncompetitive. Wealth shifts to the inventors of digital. Fossil fuels face an uncertain future, with digital running rampant, destroying demand and unlocking reserves. Renewables increasingly behave like digital and continue to eat the energy world.

No one of these four is a "correct" scenario, of course. There are many possible futures, as there are many trend lines that could be considered. But the trend lines above do not show any signs of abating, so a thoughtful response is to watch how the uncertainties play out and to manage accordingly. For example, fuel retailers may wish to start small trials of new vehicle-servicing models in select markets to measure customer interest and impact, before embarking on a whole-of-portfolio reinvention. Data services should consider how they might grow their business, should host governments suddenly restrict data from leaving the country.

WHAT IS DIGITAL STRATEGY?

Digital strategy is the set of reinforcing choices that direct an organization's actions towards integrating digital technologies into its business. These choices should be informed by: a clear-eyed view of the goals of the business, its competitive position in its markets, the problems the company is trying to solve, the digital maturity of the industry and its ecosystem, the competitive gap to be addressed, the level of digital enablement of critical processes, and an understanding of the organization's broader journey towards becoming more technologically enabled (digital efforts should not be undertaken in a vacuum).

For example, a digital strategy for an upstream company should be one that makes a material difference to its key variables for success (growth in reserves faster than growth in production, efficiently producing oil, an efficient use of capital, and selling oil close to the market price); such a digital strategy will be strongly aligned with, and supporting, the business strategy. Of course, the business strategy will be flavored by the nature of the resource, reserves location, extraction methods, capital availability, venture structures, and supply-chain capabilities, to name a few factors, and the digital strategy needs to take all this into account.

For many companies, whose goals may be to keep pace with a peer group, integrating new technologies into the business, and capturing productivity gains as they go, may be sufficient. However, Boards everywhere should be alert to the possibility that one of their competitors (or, more than likely, some unanticipated player) is busily creating a new business model based on digital technologies that could transform the industry. This is already happening in downstream petroleum retailing, with the fuel delivery to vehicles (potentially stranding the retail asset) and mobility as a service (which eliminates car ownership as a social need, and again potentially stranding retailing).

In addition, companies need to be mindful of the nuances of each technology's return on investment, factoring in not just cost, but also time and effort to implement. Some digital ideas, like apps on smartphones for field-workers, pay back very quickly, whereas cleaning up a messy data environment for autonomous equipment may take much longer. In the end, digital strategy will help an organization realize its strategic goals by enabling it to make consistent, well-informed, and balanced decisions regarding the integration of technologies across the various dimensions of the organization.

Why Set a Digital Strategy?

The oil and gas industry knows that it needs to step up its digital game, but with multiple possible future scenarios, and a number of choices to be made, companies need digital strategies. Clearly, leaving digital to chance did not help retailing, media, entertainment, taxis, or banking. The incumbents in those industries have struggled for years as new entrants gain ground. Running a portfolio of small experiments as a strategy may help gain deeper understanding of how digital technologies behave—but at a significant cost if a competitor (either a current one or an unknown entrant) is busy building the next business model. Unfortunately, the seductiveness of digital leads many companies to conclude that any investment will make some economic sense (after all, most business processes can stand a little improvement).

Digital strategy is complicated by the fact that these technologies evolve at a much faster rate than oil and gas infrastructure. Indeed, much oil and gas infrastructure was never intended to evolve. Anticipating where digital innovations will head could be a fool's errand. Historically, it's been easy to find the latest innovation in oil and gas—there are well-established forums, trade shows, and publications for sharing valuable innovation. Digital innovation is now occurring anywhere, which makes gaining a line of sight on impactful changes much harder.

To illustrate how challenging digital choices can be to an oil and gas company, consider a midstream operator's choice between investing in a fleet of drones or enabling operations with artificial intelligence. The drones could automate the monitoring of assets and improve maintenance scheduling (a proven use case in solar plants, offshore oil assets, and mines), whereas the AI solution could provide operators with a one-hour look-ahead on asset performance and support better decision-making. These are two highly impactful but very different options, and if an investment in only one is financially feasible, how should that choice be weighed? Clearly, the "right" choice would depend on how the company views future scenarios unfolding, the alignment with organizational goals, the respective value propositions, as well as their relative cost and effort to implement.

Digital innovation can also lead to surprising new uses and markets. For example, a petroleum distribution company developed a low-cost sensor, cloud connection, and volume dashboard for its own fuel business, only to discover that its innovation was highly sought after by its own suppliers as well as customers. It now sells this innovation as a market solution. A field service company developed its own cloud-based dispatch system for its trucks, but after the downturn came (and the trucks were sold off), it began to sell its digital innovation as a going concern.

Some companies will wonder if this is a good time to invest in digital innovation. Scenario thinking informs timing, of course, but timing is always tricky. As commodity prices decline, capital becomes scarce, which causes a pull-back in technology spend; as prices rise, margins expand, which tends to promote capacity expansion rather than technology spend. The period of low commodity prices beginning in 2014 and continuing to 2018 created the right conditions for digital investments because the mantras of "lower for longer" and "lower forever" neutralized the arguments to simply wait out the downturn. First movers with digital have discovered that these technologies allow for more cost-effective scaling of activities than the more manual processes of the past, and they can satisfy a greater range of purpose.

Companies are also discovering that digital technologies provide more value through better scaling of output-to-costs than historically possible. In other words, digital technologies will help to keep costs low, even when there is an expansion of productive capacity in operations (and the potential to return to the inflated cost levels of earlier times, a concern I often hear).

However, these technologies do require significant investments to implement and the relative value of each technology will vary across different organizations. So, for oil and gas leaders that are being tasked with achieving the highest return possible on limited-capital budgets, there is a need to assess where efforts are best served and investments are best placed.

Adapting Current Business Strategy for Digital

It is no longer plausible to think of digital strategy as independent from business strategy. As I have argued throughout this book, digital innovation applies to all aspects of the business, both to the operational work of making things and moving things, and to the commercial work of financing and selling things. It enables new business models, new asset classes, and entirely new industries, which by definition means digital *is* someone's business strategy. So, to view digital as a strategy bolt-on is to be willfully blind to the potential of an unexpected business model suddenly emerging. Companies that have held to this perception have regretted it.

Unfortunately, there are dozens of business strategy methods out there, most of which predate the rise of the digital industry. Chances are good that if you're reading this book and you're presently working in the oil and gas industry, your company has adopted a strategy method, and it probably has no obvious reference to digital in it. Changing that method is going to be a painful challenge—the method and its output are likely tied to the financials, have links to the capital plan, and are reinforced through personal performance metrics. If swapping out the strategic planning process is not tenable, then a good alternative approach is to build digital thinking into the existing method.

Changes to your existing strategy approach will help your company move ahead.

- Develop a handful of possible scenarios to help surface the unknowns that inform views of the future.
- Set some tests and targets for all business units (including brownfield plants, line departments, and support functions) for their digital content. For example, business units could be mandated for an arbitrary 5 percent digital investment target in their plans.
- In more demanding settings, set targets that are reflective of the kinds of exponential improvements that are typical of digital—say, 40 to 50 percent improvement in productivity or cost reduction, instead of the usual 5 percent.
- Re-baseline the knowledge of digital across the organization, so that the contributors to the strategy think about digital in planning. Running internal trade shows and demos will help raise awareness of digital's impact.
- Update the corporate risk register to include digital change as a risk. Business model risks triggered by digital should be both high impact and high probability.

- Task the team that assembles the strategic plan to survey the market for potential business model threats to protect against disruption. Feed that insight into the planning cycle.
- Incorporate a view of how competitors, customers, regulators, and key stakeholders are including digital change in their businesses.
- Set aside some investments for improving key digital foundations, which include the data environment, ERP, and cyber. These cross-business areas won't likely surface in individual business unit plans.

Embedding Digital into Business Strategy

If your organization has concluded that it is time to move to a new business strategy method, or if it lacks a method in the first place, a good one to consider is the Strategy Choice Cascade, developed by University of Toronto professor Roger Martin. The Cascade is widely used, well founded, and works for many organizations regardless of scale, industry, or geography. While it does not explicitly call out digital, it challenges strategic planners to consider how the company will compete, which includes digital elements, and any digital content in the resulting strategy will be well grounded and aligned with the business strategy.

The Cascade is structured as five different question topics, whose answers reinforce each other:

1. **What are your goals and aspirations?** Goals might be expressed in financial terms, production levels, or market positioning. To be the leading pipeline operator in the Marcellus, for example.
2. **Where will you play competitively?** Organizations have to choose the combination of markets, products, and services that will achieve the goals.
3. **How will you win in your chosen markets?** This could be expressed in terms such as lowest cost, best features, highest quality, best experience, deepest relationships, or best technology.
4. **What organizational capabilities need to be in place?** Capabilities are the human talent, organizational structure, or special processes that must be differentiated to win.

5. **What reinforcing systems and processes are needed?** This could be an organization-wide data lake, a new ERP system, or a refreshed compensation structure.

The most robust strategies are those in which the "how to win" reinforces the "where to play" and vice versa. The most satisfying strategies are those in which the "where to play" and "how to win" choices have the potential to meet the desired goals and aspirations. The most sustainable strategies are those in which the "where to play" and "how to win" choices are buttressed by appropriate, distinct, and reinforcing capabilities, organizational systems, and initiative programs.

Case Study #10: An Upstream Company

To show how digital thinking could be incorporated into business strategy, consider the situation of a hypothetical upstream producer with thousands of small operating wells, liquids-focused, and with ample infrastructure. Capital markets are constrained, so obtaining capital will be difficult, placing a premium on improving cash flow from operations. This company has not invested in new ways of working since the high oil prices from 2010 to 2014 pushed all energies to maximize production, but from 2014 to 2018, there was no capital to invest. Its Board has challenged the management team to incorporate more digital ways of working.

It has determined that the value of the business is strongly correlated (70 percent) to just two variables: how much of the commodity it has in reserves, multiplied by today's price of the commodity. Thirty percent is for everything else—how efficiently resources are produced and how effective the company is at replacing and growing its reserves.

It aspires to maximize production in basins it deeply understands, while minimizing unplanned well downtime. The company believes that the basin in which it is located is advantaged with ample nearby infrastructure, where a few competitors have abandoned the business; there are lots of infill drilling possibilities; and the basin is blanketed with low-cost telecoms services. It plays where it can be seen by the market as an innovation leader in its basin. It aims to win by holding well-delivery costs as low as possible, keeping unplanned downtime as infrequent as possible, and by boosting recovery per well.

In addition to building capabilities in managing its supply chain, and incorporating new low-impact drilling techniques for its infield plays, it has identified a number of digital capabilities that would be very beneficial to achieving its vision.

Its first digital capability is to enhance data integration during capital execution and handover. Much of the data about its upstream assets originates with suppliers, and improved data management about assets is probably the single most impactful investment in digital that it could make. The benefits last for the life of asset and enable many other innovations. It intends to build better and deeper data integration solutions between its capital asset and operations teams, which currently are in silos.

The second key capability, analytics (which includes data visualization tools, predictive analytics, machine learning, deep learning, and artificial intelligence) will help boost and maintain production at industry-leading levels and assist with pinpointing infill-drilling sweet spots. Applied to its reserves understanding, analytics will grow reserves. The company will run a crowdsourcing trial to see if it can benefit from a bigger and deeper pool of analytics know-how.

Its third capability is better real-time data from the field, which will help keep production up and return equipment to production faster. Surveillance and monitoring systems, such as cameras on the equipment, will save hours of travel time and improve health and safety performance. Sensors on critical production equipment like pumps and compressors backhauled to a control room will improve response time on issues and decisions about assets. This sensor data will go hand in glove with the analytics capabilities (since more data improves the returns on analytics).

Its fourth and final key new capability is in workflow collaboration. Using cloud computing, it will deploy a new collaboration work environment for its suppliers that automatically carries out callouts, assigns work orders, and manages workflow changes based on the weather. It will help deliver new wells into service much quicker than before and will return wells to service in a more proactive manner.

To manage its digitally enabled business, the company will set up an innovation center that runs small-scale experiments on additive manufacturing for spare parts for downhole pumps, as well as robotics in its finance function to explore the automation of royalty calculations.

For this strategy to work, it can no longer afford two independent teams both setting up and maintaining digital technologies, one in finance and the other in operations. It will need to create a single digital function that includes its new investment focus, existing operations support, and commercial IT function. This new organization will provide the agility that the company needs to quickly embrace digital solutions while assuring the sustainability of the existing business.

Case Study #11: Services Company

What could the digital strategy be for a hypothetical oil and gas services company, exposed and buffeted by downward cost and productivity pressures? Services companies are actually going to be more impacted by digital innovation than their customers. After all, it only takes one service company to innovate even a bit, and the next procurement may favor that supplier. The buyer doesn't even have to take advantage of digital solutions to demand digital innovation of its supply chain.

Much as with the upstream companies, the boom-bust cycle has eroded the innovation index in the supply chain. The trough is all about capital constraint, price reductions, idling equipment, and crews, and there are scant dollars to pursue digital innovation. However, the company's market scan shows several emerging business models that could disrupt the services sector, and they are monitoring them carefully:

- Uber for the field: Companies could apply cloud computing, smart devices, and equipment sensors to create new procurement models and collaborative business models for field services companies.
- Cloud-based equipment monitoring: Companies could upload downhole pump-operating data to an analytic engine in the cloud in near real time to allow for predictive pump maintenance regimes across multiple operators and suppliers.
- Inventory pooling: Technicians could convert specific spare parts inventories to pooled inventories across multiple holdings to lower the overall cost of inventory holdings.
- Virtual logistics: Logisticians could pool logistics operations across multiple players in a specific field to reduce less-than truckloads, lower carbon emissions, and reduce driving incidents.

- Integrated operations: OT could create a single integrated services company to aggregate multiple services and logistics operations in a single basin to reduce service costs.

Through the downturn, this hypothetical services company shrunk both head count and unit pay for its people. Prices are off by 20 percent, and procurement buyers are still holding the line on price. Volumes are still muted in its geographic market. Its aspiration is to secure its markets and build to match its economic performance at the market peak. It aims to maintain its high safety performance, grow into more attractive markets by acquisition, improve its margins under the current pricing model, and raise the utilization rate of its equipment to boost return on assets. It intends to compete on the basis of being the most efficient at service delivery and most open to participate in new business models.

To do this, it will invest in a handful of new digital capabilities, beginning with mobility. All of its people will be mobile, and to improve visibility to its capacity, it will leverage employees' smartphones to know their locations. By broadcasting presence, smartphones help with man-down situations, evacuations, incident data capture, and fatigue. The company will add sensors to its mobile assets, equipment, and rentals to move away from dumb kit (that is, equipment without the ability to self-diagnose or report operating condition).

The second key capability will be to interpret all the data generated by all these sensors, which requires new smarts in edge computing, big data, and analytics. There may be a new business emerging as data analytics companies offer pure data analytics as a service to its customers, and the company risks being stuck in a narrow data vertical where it can only see its operations.

The third key capability will be in improving its ERP backbone system. The last great wave of ERP adoption among suppliers was early in the century. These warhorse systems are due for an overhaul.

As a key foundation task, data cleanup will be a new priority. Customers will want much higher data quality to feed their own business analytics, so the company will embark on a data remediation program. The days of a supplier feeding field data late to the customer on paper or having out-of-date systems that cannot easily integrate with operator databases is coming to an end.

These two examples illustrate different ways to think about and apply digital to business strategy. Management should be supervising a constantly evolving digital investment portfolio with a range of initiatives in various stages of development. There should be clarity as to which are driving improved operations, which are enabling growth, and which are innovations to the business.

THE ROLE OF ECOSYSTEMS

One of the many success factors that sets Silicon Valley's technology sector apart is the close proximity of a range of different economic participants in a relatively small geography. Proximity is important—chance encounters of different businesses with different skills, market outlook, products, and services increase the opportunity for innovation. Of course, the oil and gas world already has an ecosystem of suppliers that has evolved over the years to include engineering firms, environmental advisors, and land surveyors. In the main, this ecosystem does not focus on reinventing business activity. It focuses mostly on bringing together the requisite skills and capabilities to execute an existing process (such as drilling and completing a well) as efficiently as possible.

The ecosystem for the digital world is very different: its focus is to create a context that allows digital innovation to evolve quickly to focus on business problems and to accelerate adoption. A digital ecosystem is an explicit recognition that inventors and founders may have great ideas, but they need access to a range of business inputs, such as skills, technologies, facilities, funding, advice, and opportunities to bring those ideas into commercial use. The ecosystem brings these business inputs together in a structured way to improve the speed by which innovation can take place.

Ecosystems are also trying to raise the potential of success for digital startups. Venture capital knows that, on average, a startup has a 1 percent chance of business success. A startup that is focused on a specific, known business problem and is supported by the power of an ecosystem increases its chance of success to 30 percent. An ecosystem that works together on specific issues raises the chances of success even further.

Ecosystems form around vehicles for innovation. There are three common vehicles that propel ideas from inception to market trial.

- Universities are home to the lab, where curious scientists and researchers develop funding proposals and carry out experiments to move knowledge forward. Labs often have innovations and inventions that are on the shelf, so to speak, whose inventors (the university researchers) may be looking for the opportunity to push those ideas towards commercial success. Some really large companies might have their own labs.
- The incubator is a facility that allows inventors and founders to develop and refine their ideas and solutions to the point where they are ready for small field trials. Inventors and founders do not need to spring forth from a lab—they could be former employees from the industry who have some idea about an innovation that digital could enable and need a place to develop their solutions further. Some universities also have incubators, which are often co-working facilities housing multiple startups.
- The accelerator is a facility that allows innovations that are more developed to scale and grow quickly, often in a co-working arrangement with time limits for tenancy.

TO BOOST THE opportunity for innovation success, Barrick, a mining company, pointed its ecosystem to the fundamental challenge of eliminating human workers from underground mining. A fully automated underground mine would fundamentally alter mining forever. However, the technologies needed to achieve this vision did not exist—autonomous mining vehicles existed in surface mining, but there was no technology solution for the challenge of refueling underground mining equipment. Barrick's challenge to the ecosystem was solve this task. The ecosystem, with laser focus on the problem, immediately began to accelerate innovations in battery and rapid recharging technologies for heavy equipment.

A high-performance ecosystem will have the active participation of many parts of the economy that take an interest in innovation.

- **Oil and gas companies.** The most important participants for digital innovation in oil and gas are large companies from the industry. Their active leadership (providing funding, executive sponsorship, freed-up resources, homes for running trials, and access to data) will point the ecosystem towards the biggest problems to solve and identify the largest opportunities for inventors and founders to chase. Without this demand pull, the rest of the ecosystem participants, who principally supply other necessary resources, will not quickly engage.
- **Banks.** Banks want to establish business relationships with startups knowing that some of them will grow into meaningful businesses. They offer advice on structuring bank services to help with financing activity. I have also observed that some banks are themselves working to become more digital and may have digital capabilities to combine with startups. For example, a bank in Alberta is aggressively transforming into a digital bank and is able to participate in creative digitally enabled business processes involving blockchain.
- **Lawyers.** Startups need structural advice on ownership and share structure, Board setup, taxation, and in some cases, such as blockchain startups, specific insight into the tax treatment of initial coin and token offerings. Legal advice is also key in dealing with the rapidly evolving regulatory world of privacy.
- **Scientists.** It might not be obvious, but some digital solutions do benefit from scientific insight, particularly in areas like geology, petrochemical, math, algorithms design, and measurement. The leaders in areas like artificial intelligence and robotics are likely scientists working in university labs.
- **Engineers.** Some digital innovations will benefit from engineering help in design for manufacturing, to achieve specific standards compliance, or to hit some performance targets. Engineers contribute to the ecosystem through their advisory help.
- **Venture capitalists.** Nothing happens without money, and venture capital firms concentrate pools of funds to target innovation. There

are many kinds of venture funds, which may be willing to invest as little as $100,000 and up to $5 million, to help a startup take an idea forward. Frequently, venture capitalists in specific basins will collaborate to share risk.

- **Entrepreneurs.** The experience of successful entrepreneurs is very valuable to startups. If inventors have a natural weak spot, it is probably to want to work on their solution to the exclusion of raising funds, finding staff, and selling trials. Entrepreneurs can help inventors and founders stay focused on executing the most important tasks that will be key to their success.
- **Universities.** Educational institutions play multiple roles, from providing useful space for gatherings and labs for experiments, running meet-ups, conferences, and competitions, and, perhaps most important, providing student horsepower to bring to bear on specific business needs of the digital startups. Business students can carry out market research, business modeling, and financial modeling, usually at low or no cost. Engineering students can conduct research into industry standards for digital devices.
- **Technology suppliers.** Tech companies are very quick to offer up trial versions of their own commercial digital platforms, including cloud computing, developer kits, blockchain, and augmented reality, in the hopes that a successful digital innovation using their technology will eventually drive revenue growth. Tech companies even operate their own incubators and accelerators.
- **Consultancies.** Inventors and founders can often access very valuable free insight from the Internet, but frequently that insight is not sufficient to solve specific problems. They discover that they need to contract with the specialist at commercial rates to move forward. As part of an ecosystem, consultancies can offer some very specialized pro bono advice to help a promising startup.

Left on their own, digital ecosystems will not naturally take aim at the oil and gas industry's opportunities. Based on my own exposure to the ecosystems in my home city of Calgary, which is dominated by the oil and gas industry, only a fraction of the many startups I encounter are focused on this industry. Only by taking an active leadership role in the ecosystem will the system respond to oil and gas.

Digital technologies open up new ways to access analytic capabilities and capacity, yet are underutilized by the oil and gas industry. Math and data sciences are steadily advancing, and solving formerly impossible problems, but this know-how might not be readily available to the oil and gas sector, which tends to rely on a smaller number of suppliers to the industry.

I'm finding the shelf life of ideas, practices, and computations to be falling very rapidly, and these new good ideas are not originating exclusively in oil and gas. The reach of the Internet, the spread of knowledge, and falling compute costs mean that oil and gas problems could be tackled through crowdsourcing. Examples include scheduling and routing optimization, the design of joints in steam piping, or the identification of sweet spots for drilling and resource extraction.

The more adventurous oil and gas companies are prepared to supply some of their data assets, perhaps camouflaged for confidentiality reasons, to global crowdsourcing platforms like Kaggle. Through Kaggle, companies can crowdsource improvements to their existing analytic algorithms, run hackathons to create solutions to problems, and use competitions to access expensive or hard-to-acquire analytic know-how.

SETTING GOALS FOR DIGITAL

One of the most important actions for Boards and management is to set meaningful goals for their organizations, both to inspire and motivate. Depending on cultural norms, management may set goals for the business year, which Boards then endorse, whereas other Boards may be a little more directive with management. Certainly, shareholders should expect Boards to challenge management to make sure that managers are not myopically focused on the activities of the day and oblivious to the changes happening around them.

Unfortunately, this is all too common, particularly in the upstream segment of oil and gas, where change happens very slowly. It takes years of patient work to move an innovation from the lab bench to one or

"Our joint success in modernizing our operations at Texmark Chemicals has been possible because of the people who make up the ecosystem of vendors, partners, operators, and advisors working together. Each party checks their company badge at the gate and works towards the team's common goals for the Refinery of the Future, bringing together skill sets that would not otherwise reside in one place."

DOUG SMITH, CEO, TEXMARK CHEMICALS, INC.

two pilot sites that run for a couple of seasons to commercialization in the global marketplace. In one of its clever sector-comparison studies, McKinsey estimated that it takes typically thirty years from idea inception to 50 percent market penetration for an innovation in oil and gas. Compare this to five years or less for the telecommunications industry.

Developing the Framework

A maturity model is a handy tool for assessing how mature your company is on the various dimensions of digital capability that are important for your segment of the industry and for gauging where you need to be for competitive purposes. The gap between the two (current position and desired) helps with setting appropriate goals, such as moving from unplanned and random digital trials to a more thoughtful digital strategy.

Goals for a digital future should not be randomly generated; they should be cohesive and mutually reinforcing. I like to use a four-part framework with Boards to stimulate their thinking on appropriate goals for management.

1. BUSINESS MODEL

The biggest worry for Boards is that some new business model will pop up from nowhere to displace the incumbents before they have enough time to react. Recall that one company's margin is another company's meal: Which parts of the business model look ripe for targeting? Business-model pressures have emerged already in retail and mobility, but there is also opportunity in:

- construction, where margin on margin is common;
- upstream data, where cloud computing could cluster data more effectively;
- field services, where collaboration could optimize service delivery;
- carbon management, where blockchain could be used to track carbon emissions;
- refining and chemicals, where IP and value lies in owning the "recipe," not necessarily the assets of production; and
- financing, using cryptocurrency to transform funding and fractional ownership.

2. DIGITAL DISRUPTORS

With digital racing off madly in all directions, it's not entirely clear which digital innovations look like they will have the biggest impact on oil and gas. Waiting for the future to arrive is one tactic, but another would be to monitor those that appear to have taken up already. To recap, these include:

- cloud computing, which is key to leveraging all the distributed compute power;
- artificial intelligence, analytics, and machine learning, which enable better business decisions;
- augmented reality, which enables deeper understanding of assets and work;
- the Internet of Things, which puts all machines on the network;
- autonomous technology, which automates human routine work via robots;
- additive manufacturing, which transforms supply chains and maintenance; and
- blockchain, which is fundamental to transforming business relationships.

3. DIGITAL FOUNDATIONS

There are four foundations that need to be in place for digital innovation to be fully successful. I reckon most companies in oil and gas have some of these foundations already, but they may need shoring up to support a digital future. The foundations are:

1. ERP: the big backbone systems which are themselves embracing digital.
2. Data: the feedstock for digital innovation.
3. Cyber security: more digital sensors means more attack points to be reinforced.
4. Infrastructure: enabling all the devices and cloud computing requires better telecoms infrastructure.

4. PEOPLE

Don't tell anyone, but technology is actually the easy part of digital. The hardest part by far is the soft stuff—securing the talent to drive digital, creating fertile conditions for digital adoption in the field, training the workforce on digital concepts, rewarding early adopters who show the way. This dimension of the framework looks at how companies are approaching talent for digital.

Putting the Framework to Work

In the annual business-planning process, management sets out its goals for the year. These include the usual elements—health and safety objectives, expansion and growth spend, optimization and margin improvement, acquisitions and divestitures, people and talent targets, social outcomes, financial commitments. What's generally not included are specific goals related to digital.

I would add the above framework to the planning cycle to provoke management into thinking about digitally enabled change. The framework needs to be pushed down the organization—it won't do to have just a handful of millennials in the home office dreaming up new uses for blockchain. It has to be real for the front line, too. Ideally, the front line pulls digital into its business, rather than head office pushing digital onto the front line. With management now embedding digital goals into the plan, the Board can then play its proper role to review and challenge management in the planning cycle.

Boards should be looking for these kinds of goals that signal that digital is being treated seriously:

- Management is messaging to the organization and markets about the importance of digital for the future.
- Front-line employees are engaged, being shown, not told, how digital is impacting the industry.
- A digital strategy is being developed for the organization.
- Measurable business improvements will be attributable to specific digital change.
- Funding is allocated for a portfolio of small digital experiments in key areas.

- Management is regularly monitoring emerging and disruptive businesses.
- There is investment in the foundation (ERP, cyber, data, infrastructure) with an eye to the future.

By embedding the framework into the planning cycle, management acknowledges that digital is not merely a once-and-done, but that it is a journey. That journey will take many years to complete, if it ever comes to an end. Scenarios of the future give some sense as to its general direction, and knowing where you are starting from means that you can set more realistic and achievable goals.

A USEFUL WAY to view goal-setting is to position your company on a digital maturity matrix. This tool will provide multiple levels of maturity across different digital elements. Each company should decide, according to its strategy, where it wants to operate across each element. In some areas, a strategy may dictate leadership; in others, a fast follower. Annual reviews of the maturity model help track progress towards company goals.

In time, digital will not be something that needs to be added to the management agenda—it will be deeply embedded in how management thinks about the business.

THE ECONOMIC ARGUMENT

Oil executives are a skeptical lot. In late 2017, one of Calgary's leading oil sector CEOs asked about the business case for digital in the upstream: "Bring me a $3-per-barrel savings from digital and you have my attention." With the goal now set, and after several years of cost-cutting and survival strategies, could there remain some big nuts to crack, and could digital innovation help? To find the $3, let's assume the production cost per barrel is $25, so we need to find a little better than 10 percent on a production volume of 1 million barrels per day. Assume 25 percent of that is labor ($6), 20 percent is energy ($5), 30 percent is capital ($7.50), and 15 percent is services ($3.75).

A STANDARD PRODUCTION cost of $25 per barrel is precisely wrong—no one oil company would have such a production cost across all its facilities at all times for all of its wells. Some will be much higher, others much lower. Certainly, for the higher cost producing assets, like oil sands, $25 would be directionally correct and might well be optimistic.

Confronting the Orthodoxies

The industry operates on a number of founding premises, or orthodoxies, that digital innovation can now call into question. These orthodoxies provide tight boundaries to innovation in the industry, crowd out other ideas, and create similar patterns of thinking and problem-solving that constrain the business models in the sector. Here are some of the orthodoxies that the courageous and differentiated oil and gas company will seek to reinvent in a quest to change the cost structure of the industry.

I GREW UP with the orthodoxy that it was not safe to talk to strangers and to not get into cars with strangers. Now, I use my phone to summon a stranger and I get in his car, and at my destination, I leave without physical payment.

ORTHODOXY #1: PEOPLE CONTROL INFRASTRUCTURE

Throughout the entire sector, people control the assets and the infrastructure. National oil companies even serve as a kind of employment agency for the country. Everywhere you look, you find padded seats and human-centered controls, steering wheels and rearview mirrors, pedals and levers. It doesn't matter whether it's a truck, a rig, or a control room—there are always lots of people about, and people are at the center of all the activity. Oil and gas assets are not self-controlling, self-managing, or even self-diagnosing. They are, at best, self-reporting.

And oil and gas people are expensive. Oil and gas compensation is among the highest in the world in dollars per hour. As skills shortages appear in the sector, salaries for all jobs drift upwards. Some Australian offshore gas projects paid north of $200,000 per year for "laundry technicians." Worse, these high wages lift up the cost of white-collar services and the field services.

Robots in the house: $0.90 savings per barrel

On the horizon, I can now clearly see self-driving trucks, self-driving trains, and soon pilot-less helicopters. This is only the start. The idea that only a human can take a seat, read the dials, handle the controls, and act is fast becoming an obsolete idea. Robots can take the place of people in many different roles in the sector. Just in the front office, robots have made great strides—the mining industry began working with robotic trucks some ten years ago. Oil and gas can support innovations that result in the single-person drill crew, single-operator work over rigs, lights-out plants, automated tank and pipeline inspectors, robotic welding, and drone-based field supervision. A Calgary company uses robots to process royalties, at 95 percent productivity improvement over its people-centered legacy approach.

Robots take no breaks, no vacations, no training courses, and no rest. They achieve dramatically better productivity of an equivalent human in a well-defined task area. It is no surprise that China is the number-one buyer of industrial robots in the world; even with their labor-cost advantage, they see the value of robots.

Robots could displace 10 percent of labor cost in the front office and half as much again in the back office.

ORTHODOXY #2: DATA IS AN EXPENSE

If there's one thing that big digital companies have taught us, it's that data is actually an asset, and in fact it might be *the* asset. Oil companies still see data as an expense. Data is not generally reflected on the balance sheet and it isn't managed as a strategic resource. Data sits in functional silos, organized by the line departments of operations, engineering, facilities, and assets. Engineers have been trained to view "wet" memory as superior to silicon systems. There are few chief data officers to set data policies, ground rules for data collection and use, and mechanisms for sharing and analysis. One of Australia's largest gas producers estimates that their engineering team spends 40 percent of their time just tracking down the correct data for whatever analysis bedevils them at the moment.

Clean, consistent data: $1 savings per barrel

One of North America's larger oil and gas companies recognizes this opportunity and sees how poor-quality data will clog up the robots when they eventually arrive. It is removing poor-quality data (initially aiming at engineering content), correcting bad tags, and linking disparate databases together. Savings show up quickly in streamlined work planning, with fewer errors and better health and safety outcomes. The company estimated about $1 per barrel in savings over its 1-million-barrel daily production. That's before it calculated the opportunity to create a digital twin of its plant to optimize production, adopt 3-D printing to produce parts much faster, and capture savings across the supply chain.

ORTHODOXY #3: MANAGE THE BUSINESS LIKE A PROJECT

Conventional oil and gas wells were rightly treated as projects, having little in common with each other. In fact, upstream assets are more like construction projects, where every building is customized. Methods of oversight, funding, reporting, and execution followed practices from the project-management industry. Flexibility was king. When geology was more uncertain than today, and commodity prices were more robust, the higher costs associated with project execution were more than tolerable—they were immaterial.

The interface between an oil and gas company and the army of service companies that it employs hasn't materially changed in decades. It is governed by relationships and history, rather than cold industrial logic. Communications still center on phone calls and work-tracking lives in Excel.

Manage the business like a manufacturer: $0.40 savings per barrel

With a smart device in every pocket, ubiquitous networks, slick math, cloud computing, and instantaneous communications, the ability to bring manufacturing thinking to basic practices like well delivery and asset maintenance is here. Instead of random service orders sent to random services companies as random assets fail, innovative companies use big data to identify when assets are likely to fail and then schedule preventative services to carefully selected services companies. Early adopters have already exploited this opportunity and report 15 percent savings through reductions in operator unit and task pricing, optimized operators, improved quality of field services, improved production levels, fewer road miles driven, higher productivity of workers, and on and on. Think Uber but for oil and gas.

ORTHODOXY #4: ENERGY USE IS ALREADY OPTIMAL

Energy inputs in oil and gas are managed in much the same way as other industries. Contracts and procurement might negotiate supply contracts, but the management of energy consumption is highly fragmented. Each business unit manager has their energy-cost budget, and those budgets receive scant attention. Few companies, including in oil and gas, bother to make the trade-offs between different energy inputs or costs across the whole of business to optimize the full energy footprint.

Diesel is still the fuel of choice for engines, and natural gas is the fuel of choice for plants, but how can this logic hold up when the price of carbon keeps going up while the cost of renewable energy keeps falling?

Optimize energy: $0.50 savings per barrel

Google provides a graphic illustration of this point. Their engineers were certain that energy use in their data centers was optimal, and, after all, they operate the world's biggest data centers. Who could possibly do it better?

By feeding all of the compute load, machine characteristics and configuration, operating temperatures, power generation, and consumption data from each individual server and switch in the data centers into an AI engine, along with weather (the sunny side of a data center warms up faster than the shady side), energy pricing, and grid performance, the AI engine found another 20 percent energy reduction. Savings were from insights that were impossible for humans to detect using limited modeling tools like Excel.

Oil and gas producers, with their thousands of producing assets, have similar potential. AI enables the optimization of energy consumption across the fleet. Digital sensors can turn individual assets into both consumers and producers of energy, allowing more flexibility in energy management. State of the art players install solar panels and batteries on their production wells to drive pumps and facilities and use digital tools to manage their network. The truly edgy use natural gas from wells that might otherwise be flared or emitted to generate power with a small turbine, which powers up a Bitcoin-mining fleet in a sea container adjacent to the well.

ORTHODOXY #5: PAPER-BASED PROCESSES WORK WELL

The amount of paperwork that flows between players in the industry is immense: RFPs, purchase orders, requisitions, packing slips, specifications, warranties, contracts, invoices, agreements, and compliance reports. Tickets from the field are still manual, clipboards proliferate, and there's a veritable avalanche of contracting paperwork flowing up and down the supply chain. Most of these artifacts have pride of place in the finance, production accounting, supply chain, and engineering functions, enacted over the years to assure good business controls. There are even companies whose business models are to sit between the oil companies and the services companies and fix busted data before cash gets held up in invoicing.

There's a cost to all of this paper-handling. I am aware of at least one oil and gas company that employs more people to comply with water regulations than it employs in the exploration department. In a study in 2014, Deloitte Access Economics tried to calculate the cost of red tape on the Australian economy. It concluded that as much as 25 percent of the entire Australian economy is consumed by complying with controls, most of which are self-imposed by business.

Deploy blockchain with business partners: $0.50 savings per barrel

At the root of this approach to business is a lack of trust between parties. Blockchain is starting to demonstrate just how much cost could be eliminated from using technology to create high-trust systems. Two global supermajors deployed blockchain in the part of the business where they sell, ship, store, and buy petroleum products between one another. They reckon that between 30 and 50 percent of back-office processing costs could be eliminated. Some cherished business documents like invoices will become utterly superfluous.

The Bottom Line

There it is: $3.30 per barrel saved, and I think I'm just scratching the surface. We haven't talked about the construction cycle and how much improvement there could be, or how digital twin technologies could optimize plant production, or the use of augmented reality and visual analytics to raise worker performance.

The real problem facing oil and gas is actually one other orthodoxy—that the industry needs big solutions to yield big savings. This orthodoxy needs to be booted to the curb, along with all the others.

KEY MESSAGES

The role of the Board is to take a close-up view of things distant, and a distant view of things up close. Digital change requires Board attention.

1. The opportunities and the risks inherent in digital innovation will vary by business. A one-size digital answer does not fit all.
2. Boards need to be better positioned to guide their companies through the change and need to raise their digital savvy.
3. With a few adjustments, existing strategic planning can accommodate digital innovation.

4. Oil and gas needs to nurture a whole new ecosystem to be able to exploit digital innovation.
5. Companies are at risk of not being ambitious enough in an exponentially shifting world.
6. The economics of digital innovation are compelling.

CONCLUSION

IN LATE 2016, after returning to Calgary, Alberta, following a four-year stint in Brisbane, Australia, I began to study more seriously the wave of digital change that had disturbed the established order in many industries. My exploration took me into the boardrooms of many companies, governments, trade associations, startups, accelerators, financiers, think tanks, universities, conferences, and technology companies. This research turned into a series of articles published on a blog that looked into the various dimensions of the oil value chain, emerging digital technologies, and the management challenges that confronted the industry's early adopters. Just a few weeks into 2018, I began to piece the story together, like a quilter who takes fragments of leftover cloth and builds a tapestry (it looks good from a distance, but up close it's pretty ugly).

It was not obvious that digital innovation would have the same kind of impact on oil and gas as it had on entertainment, retail, financial services, and news media. After all, the industry is focused not on bits, but on molecules, and those molecules have no substitutes. How would data, analytics, and communications combine to lower costs, improve productivity, and expand resources? Skeptics, and there are many, view digital at one end of a spectrum as a solution looking for a problem and, at the other end, as another play by technology companies to get rich at their customers' expense. Advocates, and there are few, are seeing early signs that digital innovation could be highly disruptive to the industry by enabling very different business models. Close to home examples include demand-destroying, mobility-as-a-service apps that virtually eliminate private car ownership in dense urban settings; asset-destroying fuel delivery to customers' vehicles that makes retail stations obsolete; and supply-expanding machine-learning algorithms that bring high-end geologic interpretation to the low end of the market. The business case to at least look at digital solutions has become quite compelling.

I personally find it hard to keep pace with digital evolution, because I tend to think, as many in the industry, in a linear fashion. Ask any golfer to estimate thirty yards on the fairway and they can pretty reliably point to a spot in the distance, and even pick out the right club to use to land their ball. Ask that same golfer to pace out thirty exponential yards, and the task gets much harder—it's a distance about twenty-six times around Earth. Someone who has started to understand digital change and has taken even a few tentative steps will be miles ahead of the competition and, with each step, become dramatically harder to catch.

When I started this journey, I did not think that any automaker would fully follow Tesla with its electric cars, solar roofs, and batteries, but Nissan has announced an identical business model. I believed that Google Glass was a failed product, but it wasn't—it was just the first few exponential steps towards industrialized augmented reality. I considered blockchain as a tool for the underworld to purchase arms, and it is, but Walmart uses it to protect its customers from food illnesses, and Porsche will soon roll it out across its cars. I dismissed the voice assistant in my phone as incapable of understanding me, and while it still can't, IBM's Watson gives Woodside access to the power of 1,000 engineers.

I am worried for people. Today's fifty-five-year-old who monitors land agreements for a living is probably okay, but a forty-five-year-old? I cannot see that career lasting twenty more years. A thirty-five-year-old? Forget it. Vast swathes of economic activity are about to get razed by digital innovation, and most people are not prepared for the shift, are not thinking about the new skills they need, and do not have a lifetime habit of learning new capabilities. Most do not retain close ties with their alma mater institutions of learning. I used to joke that a fifty-five-year-old with thirty years of experience actually had just one year of experience repeated thirty times. Employers will need to up their retraining game to retain talent, but many are unprepared. Fortunately, universities and colleges are turning their attention to delivering the job skills of the future, but as change picks up pace, can they move fast enough?

A mindset shift is needed to move the industry to a place of readiness to embrace digital innovation, beginning with Boards. Investors are now exerting very strong pressure on Boards to address climate change and the impacts on shareholders from the shift away from fossil fuels. Digital innovation can extend the life of the existing fossil-fuel energy system by lowering its costs to compete with renewables, as well as by reducing its carbon footprint. Few technologies offer the competitive combination of lower cost, higher productivity, asset life extension, and reduced environmental impact. That's before considering the potential for entirely new and disruptive business models to emerge.

Meanwhile, the industry could wait for more data to come to light, for someone else to make the first move, for digital innovation to be fully de-risked, and for governments to create the right tax policy. But that's a strategy destined to fail. It's time to get on with it. I like Richard Branson's ethos for the Virgin Group: "Screw it, let's do it."

ACKNOWLEDGMENTS

RACHAEL AND I are indebted to the contributions of many individuals who contributed their time, talent, and insight into this important topic.

We would like to thank: Ron Brookfield, Crissy Calhoun, Ryan Cann, Bruce Conway, Dipankar Das, Dr. Ken Dick, Ian Enright, Judy Fairburn, Rony Ganon, Jean-Michel Gires, Michael Habeck, Lyon Hardgrave, David Hone, Ana Johns, Greg Lake, Aidan McColl, Magesh Pillay, Marc Pritchard, Linda Salinas, Heather Sangster, Laurel Skidmore, Andrew Slaughter, Doug Smith, Dr. Prashanth Southekal, Billy Spazante, Terry Stuart, Steve Suche, Brian Truelove, Dominika Warchol-Hann, Trena White, and Randy Wilson.

GLOSSARY OF TERMS AND ABBREVIATIONS

2G, 3G, 4G, 5G: different standards for wireless communications. 3G is about 250 times faster than 2G; 4G is 10 times faster than 3G; 5G is 10 times faster than 4G.

A

accelerators: a program of usually fixed length for a cohort of startups that features seed investments, access to experienced entrepreneurs, mentorship, training and pitch preparation.

additive manufacturing: a manufacturing process where material is sprayed from a nozzle following a design to create an object layer by layer.

adware: software that automatically displays advertising to an online user.

AFE: authorization for expenditure. A financial process in oil and gas that results in the approval of expenditures for an undertaking, such as drilling a well.

Agile: a software development technique involving iterative development where the requirements and solution evolve through collaboration between multifunctional teams.

AI: artificial intelligence. The field of computer science where computer systems perform human-like tasks, like visual perception, speech recognition, decision-making, and translation.

air gap: a device with a sensor that can process data, but lacks the ability to connect wirelessly to a network.

API: American Petroleum Institute. API gravity is a measure of how light a petroleum liquid is compared to water. Petroleum of API greater than ten floats on water. Less than ten sinks in water.

API: application programming interface. A set of reusable software (routines, protocols, and tools) to allow different systems to interact with each other.

AR: augmented reality. Computer-generated images or graphics superimposed on a view of the real world.

ATOMIC: in blockchain, a mnemonic meaning asset, trust, ownership, money, identity, and contract.

B

bbl: barrel. A common measure of oil volume. A barrel of oil is 35 gallons in the Imperial system or 42 U.S. gallons. There are about seven barrels of oil in a metric tonne. Conversion factors for oil measurement are found in many online resources, including OPEC, the IEA, and the websites of many oil companies.

big data: unstructured data in extremely large datasets.

bit: a binary representation of data. A bit is either on (represented by a positive electric charge, or the number one), or off (represented by a negative electric charge, or zero).

Bitcoin: a cryptocurrency, implemented using blockchain technology.

blending: in petroleum, the mixing of two similar but chemically different products to create a third product. High-sulfur diesel blended with zero-sulfur diesel yields low-sulfur diesel.

blockchain: an innovative technology combining distributed computing and encryption to enable trust-less business transactions.

BOE: barrel of oil equivalent. This term is shorthand that quantifies the amount of energy that is equivalent to the amount of energy in a barrel of crude oil.

BOM: bill of materials.

BP: a European oil company.

brownfield: an existing operating asset or facility.

byte: eight bits. Since each bit can be either zero or one, eight bits together provides 256 possible combinations of zeros and ones, or enough combinations to represent ten numbers (zero to nine), twenty-six letters (A to Z), and other useful symbols like decimal points, commas, and dashes.

C

canning: refers to the practice of delivering fuel in very small quantities, using a jerry can.

"cap and trade": an economic mechanism to limit greenhouse gas emissions. Cap limits the absolute amount of emissions permitted, and trade allows companies that overproduce to trade capacity to emit with those that under produce.

CH_4: methane. Its chemical composition, one carbon molecule and 4 hydrogen molecules.

closed data: private data that is not freely shared or circulated.

cloud computing: shared computer processing and data storage services available through an Internet browser.

compressor: a pump that applies pressure to a gas to reduce its volume.

conventional: refers to oil and gas wells that rely principally on natural underground pressures to force hydrocarbons to the surface.

CRIND: card reader in dispenser. Also called "pay-at-the-pump" technology.

crude slate: the choice of crude oil used in an oil refinery, and an important determinant of refinery profitability.

cyber: referring to activities carried out exclusively online.

D

data lake: a collection of datasets into a single structure to ease searching and access.

DDoS: distributed denial of service. An attack on a computer system consisting of millions of simultaneous requests for service, causing the receiving system to become overwhelmed and deny access to legitimate requests.

demurrage: a charge for detaining a transportation asset beyond an agreed time. In shipping, demurrage charges accrue when a ship is in port for loading and unloading outside the charter terms.

digital twin: a virtual version of a physical asset, used for simulation, forecasting, prediction of behavior, and training.

DLT: distributed ledger technology. Also known as blockchain.

downhole pumps: a water or oil pump that sits at the bottom of a well and pumps liquids to the surface.

downstream: activities in the oil and gas industry involving marketing and sale of refined petroleum products to end consumers, including wholesale, retail, and trading.

DR: digital reality. The collection of virtual and augmented reality tools and technologies.

drivetrain: the mechanical components of a vehicle that deliver power that enables forward movement. The dominant drivetrain in cars is the internal combustion engine and transmission system. In electric cars, the drivetrain is electric motors at each wheel.

drones: an unmanned vehicle, usually aerial, but also submersibles and vessels.

dumb technology: hardware or equipment such as pumps and motors that lack digital features such as onboard sensors and network communications.

E

ecosystem: a group of educational institutions, financiers, think tanks, suppliers, real estate facilities, entrepreneurs, and public sector agencies that together support the development of startups and growth companies.

EPC: engineering, procurement, and construction. A business offering these services.

ERP: enterprise resource planning, a kind of computer system that provides a single database and set of routines spanning multiple parts of a business, such as operations, finance, and marketing.

EV: electric vehicle.

exploration: activities in the oil and gas industry involving the search for accessible oil and gas deposits.

F

FEED: front-end engineering and design. An early design step in the engineering project life cycle that develops solutions for technical issues and estimates investment costs.

field services: services required by oil and gas facilities, typically wells, tanks, and pipelines, including inspections, repairs, and maintenance.

flat database: a kind of database design where data is stored in a simple table of rows and columns.

flow measurement: technologies that measure the volume of fluids in motion, as in a pipe.

G

G&A: general and administrative.

gamification: a set of programming techniques that exploit human predilections for competition and addiction.

Gantt chart: a project management tool that presents a work plan visually as a series of horizontal bars corresponding to task duration.

gen set: an electricity generator that uses diesel or other fuel.

GHG: greenhouse gas. Certain gases like methane and carbon dioxide accumulate in the atmosphere, trapping heat and slowly raising the earth's temperature.

greenfield: a planned operating asset or facility that is not yet in production.

H

hurdle rate: a performance target, usually financial, that an investment must achieve to be approved.

HYSYS: a software product for simulating process manufacturing.

I

IBM Watson: the collection of artificial intelligence tools and technologies from IBM.

ICE: internal combustion engine.

incubator: a company or agency that helps new and startup companies develop by providing services such as management support and office space.

infill drilling: the practice in oil and gas to locate a new well in close proximity to existing wells, to take advantage of existing facilities, like spare tankage and pipeline.

IOT: the Internet of Things. The interconnection via the Internet of devices that have some computational capability, which allows those devices to exchange data.

IT: information technology. Also the organization that supports commercial IT, such as email service and ERP.

K

Kaggle: a crowdsourcing platform for data science projects.

L

large data: well-structured data in extremely large datasets.

LIDAR: a technology for measuring distance by illuminating a target with pulsed laser light and measuring the reflected pulses with a sensor.

linear program: a mathematical model of an oil refinery that computes yields, capacities, energy needs, and blending, taking into account physical constraints of the refinery.

LNG: liquefied natural gas. "Natural" means the hydrocarbon (methane, ethane, propane, butane) is a gas at room temperature and pressure. Gases can convert into a liquid state if chilled or injected into a pressurized container. Methane converts into a liquid at -162 degrees centigrade, for example.

LTE: long-term evolution. The name of next generation wireless transmission protocol.

LTO: light tight oil. Shorthand for very low viscosity oil found in certain kinds of shale rock.

M

metadata: data about data. For example, metadata about a photo includes the date and location of the photo and the camera used to take the photo.

midstream: activities in the oil and gas industry involving the shipping of crude oil and raw gas to refining facilities and the refining of those raw materials into valuable.

MoC: Management of Change. A procedure in oil and gas facility management for introducing changes to facilities in a safe manner.

moon shot: slang for a high-risk, high-reward investment.

Moore's law: an observation by Gordon Moore, founder of Intel, that the density of transistors on a computer chip appears to double every eighteen months.

MR: mixed reality. Real world and virtual worlds merged to produce new environment where physical and digital objects can interact.

MVP: minimally viable product. A version of a software solution that offers enough benefit to motivate potential users to try it.

O

oil basin: geologic formations that feature a concentration of hydrocarbons.

oil pigs/pigging: an inspection device inserted into a pipeline that travels the length of the pipe recording inspection data (fractures, corrosion, damage, and sediment buildup).

open data: data in the public domain that can be freely used by third parties.

Oracle: an enterprise resource-planning (ERP) system.

OT: operational technology. Also the organization that supports operations technologies, such as SCADA and sensors.

P

permeability: the ability of a liquid to flow through a rock using cracks and channels in the rock.

porosity: the spaces in rock that contain oil and gas.

Primavera P6: a capital project-planning software product.

process manufacturing: a technique for producing goods in a continuous nonstop fashion. Chemical manufacturing, paper-making, and oil refining are examples of process manufacturing.

R

ransomware: a kind of computer virus that locks access to a system or threatens to destroy a system unless a ransom is paid.

relational database: a database design where data is stored in multiple tables that relate to each other through common keys or identifiers.

risk matrix: a tool used in risk management that presents risks to a business in terms of the scale of impact and the probability of those risks being manifest.

RPA: robotic process automation. A set of software tools and protocols for automating routine tasks.

S

S4/HANA: a version of SAP that has been designed to take advantage of digital innovations like cloud computing.

SAGD: steam-assisted gravity drainage. An oil-extraction technique in oil sands involving the injection of steam into an oil sands deposit for the purposes of lowering the viscosity of the oil so that it can flow to the surface through a drainage pipe.

SAP: an enterprise resource-planning (ERP) system.

SCADA: supervisory control and data acquisition. A kind of computer system that monitors operating equipment and collects data about equipment performance, such as temperature, throughput, levels and energy consumption.

smart technology: hardware or equipment such as pumps and motors that incorporate digital technologies.

stage gate: an engineering methodology that features the design and engineering of an asset in a progressive stepwise fashion, with periodic gates to approve or suspend work during development.

Strategy Choice Cascade: a strategy method developed by Roger Martin, former dean of the Rotman School of Management at the University of Toronto.

submersibles: an unmanned vehicle that operates underwater.

T

TAN: total acid number. A measure of acidic compounds in crude oil.

Trojan: a software device that enables access to a system without passing through normal security controls.

turnarounds: the practice of shutting down an operating facility to undertake repairs and maintenance.

U

UI: user interface.

unconventional: refers to oil and gas deposits that are extracted using techniques other than through conventional wells.

upstream: oil and gas industry activities involving exploring for and extracting resources and aggregating them for shipping to oil refineries.

visualization: techniques and technologies for presenting information in ways that enable understanding and analysis.

V

VR: virtual reality. A computer-generated simulation of a 3-D image or environment.

W

waterfall method: a software development technique that features the development and approval of requirements and specifications for new solutions in a stepwise fashion.

wetware: human brain power, distinct from software (code) and hardware (chips).

REFERENCES

CHAPTER 1

IOR. "Hydip Fuel Monitoring System": http://ior.com.au/fuel/fuel-management-solutions/hydip-fuel-monitoring-system.

Lee Matthews. "Criminals Hacked a Fish Tank to Steal Data from a Casino," Forbes.com (July 27, 2017): https://www.forbes.com/sites/leemathews/2017/07/27/criminals-hacked-a-fish-tank-to-steal-data-from-a-casino.

SEMI News. "Smart Sensor Sales to Double to 30 Billion Units between 2015 and 2020—But Sensor Industry Plagued by Falling Prices," Roland Berger press release (February 23, 2017): http://www.semi.org/en/press/smart-sensor-sales-double-30-billion-units-between-2015-and-2020-%E2%80%93-sensor-industry-plagued.

Government of Canada. "Applicants—Auction for residual spectrum licences in the 700 MHz, 2500 MHz, 2300 MHz and PCS-G Bands": http://www.ic.gc.ca/eic/site/smt-gst.nsf/eng/sf11390.html.

CHAPTER 2

Digital McKinsey. "Reborn in the Cloud," McKinsey.com (July 2015): https://www.mckinsey.com/business-functions/digital-mckinsey/our-insights/reborn-in-the-cloud.

SAP S/4HANA. "System Requirements": https://www.sap.com/canada/products/s4hana-erp/technical-information.html.

Google. "Quick, Draw!": https://quickdraw.withgoogle.com.

Deloitte Nederland. "Product Demo: Artificial Intelligence in the Tax Function," YouTube.com (August 17, 2017): https://youtu.be/e7Dl7AMyEoA.

IBM. "70 Miles from Shore with Watson: Woodside Energy and IBM," YouTube.com (August 22, 2016): https://youtu.be/GFZ2IaTVkY8.

Bloomberg News. "The Great Crew Change: Lost Generation of Oil Workers Leaves Few Options for Next Boom," Business. FinancialPost.com (July 7, 2016): https://financialpost.com/commodities/energy/the-great-crew-change-lost-generation-of-oil-workers-leaves-few-options-for-next-boom.

IOR. "Hydip Fuel Monitoring System": http://ior.com.au/fuel/fuel-management-solutions/hydip-fuel-monitoring-system.

International Energy Agency. "Digitalization and Energy 2017," IEA.org (November 5, 2017): http://www.iea.org/digital.
National Institute for Aviation Research (NAIR), Wichita State University. "Robotics and Automation": https://www.niar.wichita.edu/researchlabs/robotics_overview.asp.
Rio Tinto. "Mine of the FutureTM": https://www.riotinto.com/australia/pilbara/mine-of-the-future-9603.aspx.
Business of Apps. "Pokémon GO Revenue and Usage Statistics (2017)": http://www.businessofapps.com/data/pokemon-go-statistics.
Shell Global. "A Bionic Inspector Rolls In": https://www.shell.com/inside-energy/a-bionic-inspector-rolls-in.html.
FieldBit. "Enterprise Solution": https://www.fieldbit.net.
The Linde Group. "Smart Glasses: Linde Looks to the Future": http://www.linde-gas.com/en/whats-happening/smart-glasses-linde-looks-to-the-future.
Gravity Jack. "Augmented Reality in the Oil and Gas Industry": https://gravityjack.com/news/augmented-reality-oil-gas-industry.
James Pozzi. "The Impact—and Importance—of Virtual and Augmented Reality in Aviation," MRO-Network.com (June 14, 2016): https://www.mro-network.com/emerging-technology/impact-and-importance-virtual-and-augmented-reality-aviation.
Norma Kamali: http://normakamali3d.com.
Yu-Kai Chou. "The 10 best eCommerce and Shopping Examples That Use Gamification": https://yukaichou.com/gamification-examples/top-10-ecommerce-gamification-examples-revolutionize-shopping.
Satoshi Nakamoto. "Bitcoin: A Peer-to-Peer Electronic Cash System," Nakamoto Institute.com (October 31, 2008): https://nakamotoinstitute.org/bitcoin.
William Mougayar. *The Business Blockchain: Promise, Practice, and Application of the Next Internet Technology*. New York: John Wiley & Sons, 2016.
Sean Szymkowski. "Porsche First to Bring Blockchain to Auto Industry," motorauthority.com (March 12, 2018): https://www.motorauthority.com/news/1115654_porsche-first-to-bring-blockchain-to-auto-industry.
Globe Newswire. "GuildOne's Royalty Ledger settles first royalty contract on R3's Corda blockchain platform," globenewswirecom (February 14, 2018): https://globenewswire.com/news-release/2018/02/14/1348236/0/en/GuildOne-s-Royalty-Ledger-settles-first-royalty-contract-on-R3-s-Corda-blockchain-platform.html.

CHAPTER 3

Kevin Seals, "Ultrasound-on-a-chip supercharged with AI: The most disruptive technology in radiology?" becominghuman.ai (December 6, 2017): https://becominghuman.ai/ultrasound-on-a-chip-supercharged-with-ai-the-most-disruptive-technology-in-radiology-b2684b0421aa.
International Energy Agency. "Digitalization and Energy 2017," IEA.org (November 5, 2017): http://www.iea.org/digital.
Ann Muggeridge. "Recovery Rates, Enhanced Oil Recovery and Technological Limits," The Royal Society Publishing (January 13, 2014): https://www.ncbi.nlm.nih.gov/pmc/articles/PMC3866386.

John Kemp. "Decline Rates Will Ensure Oil Output Falls in 2016," Reuters (September 15, 2015): https://www.reuters.com/article/oil-production-kemp/column-decline-rates-will-ensure-oil-output-falls-in-2016-kemp-idUSL5N11L26U20150915.

Michelle Bellefontaine. "Poor Co-ordination and Communication Hindered Battle with Fort McMurray Wildfire, Says Leaked Report," CBC News (June 8, 2017): http://www.cbc.ca/news/canada/edmonton/fort-mcmurray-wildfire-report-warnings-recommendations-alberta-forestry-1.4152199.

IBM. "70 Miles from Shore with Watson: Woodside Energy and IBM," YouTube.com (August 22, 2016): https://youtu.be/GFZ2IaTVkY8.

Globe Newswire. "GuildOne's Royalty Ledger settles first royalty contract on R3's Corda blockchain platform," globenewswirecom (February 14, 2018): https://globenewswire.com/news-release/2018/02/14/1348236/0/en/GuildOne-s-Royalty-Ledger-settles-first-royalty-contract-on-R3-s-Corda-blockchain-platform.html.

Shell Australia. "Using Drone Technology": https://www.shell.com.au/about-us/projects-and-locations/qgc/environment/environmental-operations/using-drone-technology.html.

Fractracker Alliance. "Oil and Gas in the U.S.": https://www.fractracker.org/map/national/us-oil-gas.

ServiceNow: https://www.servicenow.com/products/application-development.html.

Synaptic AP. "End-to-End Salesforce Solutions for Field Service and Operations in Oil, Gas and Energy": https://www.synapticap.com/industries/oil-and-gas.

IronSight: https://ironsight.ca.

Payload: http://payload.com.

GE: https://www.ge.com/power.

Sun Wenyu. "China to Face Shortage of 3 Million Robot Operators by 2020," *People's Daily Online* (September 22, 2017) http://en.people.cn/n3/2017/0922/c90000-9272651.html.

BP. "Statistical Review of World Energy 2017," bp.com (June 2017): https://www.bp.com/content/dam/bp/en/corporate/pdf/energy-economics/statistical-review-2017/bp-statistical-review-of-world-energy-2017-full-report.pdf.

Emma Graney. "Sturgeon Refinery Costs Balloon Again to $9.7 Billion," *Edmonton Journal* (April 10, 2018): https://edmontonjournal.com/news/politics/sturgeon-refinery-costs-balloon-again-to-9-7-billion.

Solomon Associates. "Solomon Benchmarking": https://www.solomononline.com/benchmarking.

Aspentech. "Aspen HYSYS": https://www.aspentech.com/products/engineering/aspen-hysys.

Zolaikha Strong. "How Much Copper's in That Electric Vehicle?," *EV World* (September 5, 2017): http://evworld.com/focus.cfm?cid=370.

Andrew Liptak. "Italian Tesla Drivers Set Distance Record after Driving Model S 670 Miles on a Single Charge," *The Verge* (August 6, 2017): https://www.theverge.com/2017/8/6/16104628/tesla-drivers-italy-distance-record-model-s.

Exploratorium. "Drafting," Exploratorium.edu. https://www.exploratorium.edu/cycling/aerodynamics2.html.

eRoadArlanda: https://eroadarlanda.com.

International Organization of Motor Vehicle Manufacturers. "2017 Production Statistics," oica.net: http://www.oica.net/category/production-statistics/2017-statistics.
Whim. "Mobility as a Service": https://whimapp.com.
Brian Solomon. "WeFuel Joins Crowded 'Uber For Gas' Startup Fight," Forbes.com (January 26, 2016): https://www.forbes.com/sites/briansolomon/2016/01/26/wefuel-joins-crowded-uber-for-gas-startup-fight.
International Energy Agency. "Electric Vehicles: Tracking Clean Energy Progress," iea.org (July 6, 2018): http://www.iea.org/tcep/transport/evs.
Kennedy Paul. "Top 10 Car Group Manufacturers in the World in 2016 by Sales," DriveSpark.com (October 19, 2017): https://www.drivespark.com/four-wheelers/2017/top-10-car-manufacturers-in-2016-in-the-world-020233.html.
OPEC. "Monthly Oil Market Report," OPEC (November 2015): https://www.opec.org/opec_web/static_files_project/media/downloads/publications/MOMR_November_2015.pdf.
U.S. Energy Information Administration. "Frequently Asked Questions—How Many Gallons of Gasoline and Diesel Fuel Are Made from One Barrel of Oil?" EIA.gov (June 29, 2018): https://www.eia.gov/tools/faqs/faq.php?id=327&t=9.
Deloitte. "Short of Capital? Risk of Underinvestment While Oil Price Is Lower for Longer," Deloitte.com (2016): https://www2.deloitte.com/us/en/pages/energy-and-resources/articles/lower-for-longer-risk-oil-gas-underinvestment.html.
McKinsey Global Institute. "Digital America: A Tale of the Haves and the Have-Mores," McKinsey.com (December 2015): https://www.mckinsey.com/~/media/McKinsey/Industries/High%20Tech/Our%20Insights/Digital%20America%20A%20tale%20of%20the%20haves%20and%20have%20mores/MGI%20Digital%20America_Executive%20Summary_December%202015.ashx.
World Economic Forum. "Enabling Trade: Valuing Growth Opportunities" (2013): http://www3.weforum.org/docs/WEF_SCT_EnablingTrade_Report_2013.pdf.
Elizabeth Gibney. "Self-taught AI Is Best Yet at Strategy Game GO," Nature.com (October 18, 2017): https://www.nature.com/news/self-taught-ai-is-best-yet-at-strategy-game-go-1.22858.
Real Serious Games. "Forensic Animation": http://www.realseriousgames.com.
FieldBit: https://www.fieldbit.net.
RedEye Apps: https://redeye.co.
CrunchBase. "SAP Acquisitions" (June 2018): https://www.crunchbase.com/organization/sap/acquisitions/acquisitions_list.

CHAPTER 4

Smart Cap Technologies: http://www.smartcaptech.com.
Shell Australia. "Using Drone Technology": https://www.shell.com.au/about-us/projects-and-locations/qgc/environment/environmental-operations/using-drone-technology.html.
VAKT, a digital ecosystem for physical post-trade processing: http://www.vakt.com.
International Energy Agency. "Digitalization and Energy 2017," IEA.org (November 5, 2017): http://www.iea.org/digital.
BP. "Statistical Review of World Energy 2015," BP.com (June 2015): https://www.bp.com/content/dam/bp-country/es_es/spain/documents/downloads/PDF/bp-statistical-review-of-world-energy-2015-full-report.pdf.

Woodside Energy: https://www.youtube.com/user/WoodsideEnergyLtd.
Herox. "Integra Gold Rush": https://www.herox.com/IntegraGoldRush.
Woodside Energy. "3D Printing of Parts in Oil and Gas": http://www.woodside.com.au/Working-Sustainably/Technology-and-Innovation/Pages/Technology%20and%20Innovation.aspx#.W3cWed9lDmo.
Anh (Annie) Nguyen, Brooklynn Malec, Jenna Nguyen, and Ibrahim Oshodi. *Robot Revolution... or Automation Adaptation?* Calgary, Alberta: University of Calgary, 2018.
Suncor Energy. "Suncor Energy Implements First Commercial Fleet of Autonomous Haul Trucks in the Oil Sands," Marketwired (January 30, 2018): http://www.suncor.com/newsroom/news-releases/2173961.
Rio Tinto. "First Delivery of Iron Ore with World's Largest Robot," riotinto.com (July 13, 2018): https://www.riotinto.com/media/media-releases-237_25824.aspx.
Railway/Pro. "China to Develop Driverless High-Speed Train," (March 2, 2018): https://www.railwaypro.com/wp/china-develop-driverless-high-speed-train.
Sun Wenyu. "China to Face Shortage of 3 Million Robot Operators by 2020," *People's Daily Online* (September 22, 2017): http://en.people.cn/n3/2017/0922/c90000-9272651.html.
Gerrit De Vynck. "Canada to Scrap IBM Payroll Plan Gone Awry Costing C$1 Billion," Bloomberg.com (March 1, 2018): https://www.bloomberg.com/news/articles/2018-03-01/canada-to-scrap-ibm-payroll-plan-gone-awry-costing-c-1-billion.
Joe McKendrick. "Amazon Software Releases," ZDNet (March 24, 2015): https://www.zdnet.com/article/how-amazon-handles-a-new-software-deployment-every-second.
Valerie Jones. "More than 440,000 Global Oil, Gas Jobs Lost During Downturn," Rigzone (February 17, 2017): https://www.rigzone.com/news/oil_gas/a/148548/more_than_440000_global_oil_gas_jobs_lost_during_downturn.
Lockheed Martin. "Contract for Next Generation Open Process Automation System," (February 8, 2018): https://news.lockheedmartin.com/2018-02-08-lockheed-martin-awarded-contract-for-next-generation-open-process-automation-system.

CHAPTER 5

Green Car Congress. "Berkeley Study Finds Self-Driving Electric Taxi Fleets in Manhattan Would Deliver Significant Cost and Environmental Benefits," (March 30, 2018): http://www.greencarcongress.com/2018/03/20180330-saev.html.
International Organization of Motor Vehicle Manufacturers. "2017 Production Statistics," OICA.net: http://www.oica.net/category/production-statistics/2017-statistics.
Kaggle: http://www.kaggle.com.
Richard Evans and Jim Gao. "DeepMind AI Reduces Google Data Centre Cooling Bill by 40%," Deepmind.com (July 20, 2016): https://deepmind.com/blog/deepmind-ai-reduces-google-data-centre-cooling-bill-40.
Deloitte. "Rules Eat Up $250 Billion a Year in Profit and Productivity," Deloitte.com (October 29, 2014): https://www2.deloitte.com/au/en/pages/media-releases/articles/rules-eat-up-250-billion-a-year-271014.html.

INDEX

Figures indicated by page numbers in italics

ABOUT THE AUTHORS

GEOFFREY is an independent advisor to the oil and gas industry. In the past, he was a partner with Deloitte, one of the world's largest professional services firms, where he held a variety of leadership roles for various services, industry programs, and offices, following an initial career with Imperial Oil. He has worked around the world, including China, Hong Kong, Japan, South Korea, Australia, Canada, the U.S., and the Caribbean. Throughout that time he has led hundreds of projects helping companies address critical problems and opportunities. He is a frequent speaker on the challenges and opportunities of digital innovation in oil and gas, writes a weekly blog about digital topics, and hosts a popular podcast series on this subject. He is an alumni of McGill University (B.Comm) the Ivey School of Business at the University of Western Ontario (MBA).

RACHAEL values lifelong learning and innovation and always aims to integrate these concepts into the work she does. Since joining Deloitte in 2001, she has worked as a consultant with clients across multiple industries. Her journey has taken her to work directly, or as Board volunteer, in education, banking, telecommunications, technology, and—for the last twelve years—energy. Rachael met Geoffrey in 2017 when they began collaborating to bringing digital innovation to clients across the energy value chain. She earned her undergraduate degree in psychology from the University of Pennsylvania and her MBA (with a focus on information technology and business transformation) from the Sloan School of Management at MIT.

Made in the USA
San Bernardino, CA
22 January 2019